ILLUSTRATED BY SHIHO PATE

ANIMATED SCIENCE

ROCKS AND MINERALS

WRITTEN BY JOHN FARNDON

Scholastic Press • New York

CONTENTS

FIND YOUR FAVORITE ROCK OR MINERAL

Rock up 4
Rock table 6
Mineral table 8

FIERY ROCKS 10

Granite **12**
Peridotite **14**
Dolerite **16**
Rhyolite **18**
Basalt **20**
Obsidian | Perlite | Pumice | Tuff **22**

PILED-UP ROCKS 24

Shale **26**
Sandstone **28**
Chalk | Chert | Flint **30**
Limestone **32**
Travertine | Dripstone | Tufa **34**
Coal **36**

CHANGED ROCKS 38

Hornfels **40**
Marble **42**
Gneiss **44**
Slate **46**

MAGIC MINERALS 48

Mineral chemistry 50

Realgar | Orpiment **52**
Galena | Pyrite | Cinnabar **54**

Awesome ores 56

Barite **58**
Gypsum **60**
Crocoite | Wulfenite | Monazite **62**
Turquoise **64**
Halite **66**
Fluorite **68**

Spinel **70**
Ruby **72**

Mineral detective 74
Mineral detective 76

Hematite | Magnetite | Siderite |
Limonite **78**
Bauxite **80**
Calcite **82**
Aragonite **84**
Malachite **86**

Opal **106**
Topaz **108**
Garnet **110**
Sphene | Spodumene **112**
Talc | Mica **114**
Gold | Copper **116**
Diamond | Sulfur **118**
Amber | Jet | Pearl **120**

Space rocks 122

Glossary 124
Index 126

SANDY PLANET 88

Feldspars **90**
Lazurite **92**
Quartz | Citrine | Amethyst |
Tiger's Eye **94**
Chalcedony **96**
Jade **98**
Tourmaline **100**
Beryl **102**
Emerald **104**

ROCK UP

YES, WE LIVE IN A VERY ROCKY WORLD. Earth is a rocky planet, and wherever you go, you're walking on it! Our homes, our technology, everything from a model car to a space rover is made with materials that come from rocks. Rocks might all look, well, rocky! But in this book, you'll see that every rock has an amazing story to tell. You'll get to know what makes each kind distinct so you can start identifying rocks nearby. Who knows, you may find a precious gem!

ROCKS, MINERALS, AND STONES

Rocks are mostly made either from tiny grains or crystals that fit together like a jigsaw. The grains or crystals are minerals, and every rock is made from its own special minerals' recipe.

Minerals are natural solids. They typically occur as crystals—some tiny, some so big and beautiful they make gemstones. Every mineral is made from its own unique recipe of chemicals.

Stones are chunks of rocks or minerals broken off the main mass of rock by the weather.

LOOKING FOR ROCKS If you like hunting for rocks, people might call you a rock hound. But it's probably best to use your eyes, not your nose like a dog, to sniff out the best examples! Where to start? Here are some ideas:

Outside your door:
Even a small backyard or a local park can contain lots of rocks and minerals. Just open your eyes!

Wild rocks: Beaches and streambeds are great places to find stones. You can often see bare rocks in cliffs, quarries, and caves, though only visit with an adult.

On the town: Many buildings contain concrete made from limestone, sand, and gravel. Fancy buildings are coated in marble. Bricks are actually baked clay. You'll find all these rocks in this book!

ROCK TABLE

FIERY ROCKS: IGNEOUS PP. 10–23

Igneous rocks are made when hot, molten rock inside Earth, called magma, cools and hardens. They sometimes form when magma erupts from a volcano as lava, and the lava cools and hardens into rock.

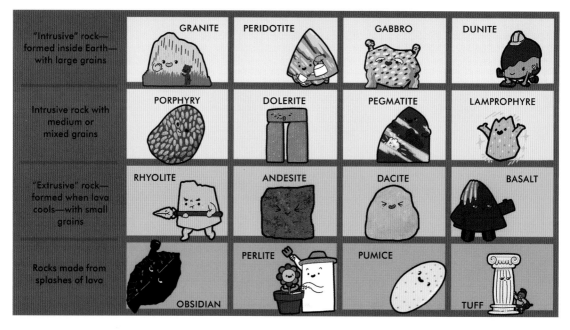

"Intrusive" rock—formed inside Earth—with large grains	GRANITE	PERIDOTITE	GABBRO	DUNITE
Intrusive rock with medium or mixed grains	PORPHYRY	DOLERITE	PEGMATITE	LAMPROPHYRE
"Extrusive" rock—formed when lava cools—with small grains	RHYOLITE	ANDESITE	DACITE	BASALT
Rocks made from splashes of lava	OBSIDIAN	PERLITE	PUMICE	TUFF

PILED-UP ROCKS: SEDIMENTARY PP. 24–37

Sedimentary rocks are made from piled-up remains of other rocks, dead plants and animals, or dissolved chemicals.

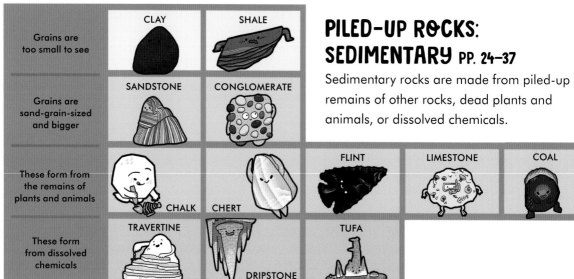

Grains are too small to see	CLAY	SHALE			
Grains are sand-grain-sized and bigger	SANDSTONE	CONGLOMERATE			
These form from the remains of plants and animals	CHALK	CHERT	FLINT	LIMESTONE	COAL
These form from dissolved chemicals	TRAVERTINE	DRIPSTONE	TUFA		

All the rocks in the world are one of three types—igneous, sedimentary, or metamorphic. Each type is divided into groups. You can see here, at a glance, all the main rock groups and where you can find them in the book.

CHANGED ROCKS: METAMORPHIC PP. 38–47

Metamorphic rocks are rocks that have been transformed by heat and pressure deep inside Earth.

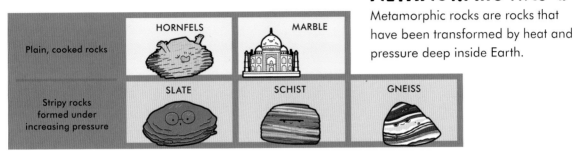

	HORNFELS	MARBLE	
Plain, cooked rocks			
Stripy rocks formed under increasing pressure	SLATE	SCHIST	GNEISS

MAGIC MINERALS: P. 48 ONWARD

Rocks contain all the thousands of kinds of minerals. But they're mostly made from just a handful of "silicate" minerals—feldspars, quartz, mica, olivine, pyroxene, and amphibole.

FELDSPARS

QUARTZ

MICA

OLIVINE

PYROXENE

AMPHIBOLE

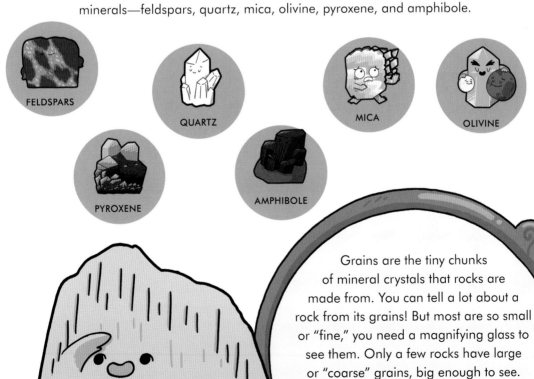

Grains are the tiny chunks of mineral crystals that rocks are made from. You can tell a lot about a rock from its grains! But most are so small or "fine," you need a magnifying glass to see them. Only a few rocks have large or "coarse" grains, big enough to see.

These form from hot liquids in Earth's crust	ORPIMENT  REALGAR	GALENA 	PYRITE
	CROCOITE	WULFENITE	MONAZITE	TURQUOISE
Minerals with oxygen and sometimes carbon	SPINEL 	RUBY HEMATITE		MAGNETITE 
Silicate minerals with oxygen and silicon	FELDSPARS	LAZURITE	QUARTZ	CITRINE
	CARNELIAN	JADE	TOURMALINE	BERYL
More silicate minerals with oxygen and silicon	EMERALD 	OPAL	TOPAZ	GARNET 
Native minerals form from only one element	GOLD 	COPPER	DIAMOND 	SULFUR
Mineral look-alikes made from the remains of plants and animals	AMBER 	JET 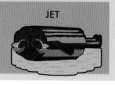	PEARL 	

Minerals are the natural crystals from which all the world's rocks are made. There are more than 5,000 different kinds of minerals, but only thirty or so are really common, and most rocks are made from just a handful. The crystals are usually tiny, and the big crystals that collectors love form only in special places.

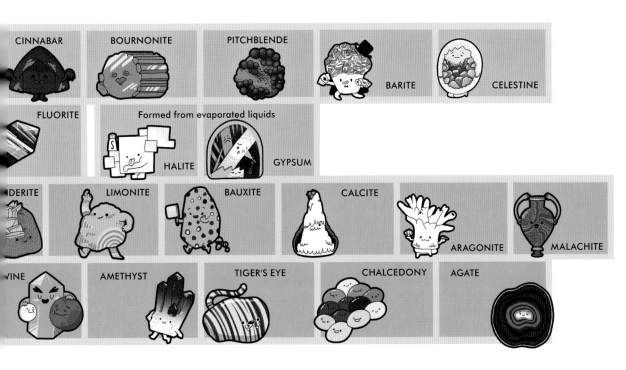

CINNABAR

BOURNONITE

PITCHBLENDE

BARITE

CELESTINE

FLUORITE

Formed from evaporated liquids

HALITE

GYPSUM

...DERITE

LIMONITE

BAUXITE

CALCITE

ARAGONITE

MALACHITE

...VINE

AMETHYST

TIGER'S EYE

CHALCEDONY

AGATE

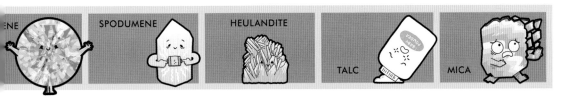

...ENE

SPODUMENE

HEULANDITE

TALC

MICA

IT'S ALL CHEMISTRY
What makes each mineral unique is its chemistry—the chemical elements from which it's made. A few minerals, including gold, are made from just one element. Most are compounds, or combinations of two or more elements. Quartz, for example, is a silicate: a compound of oxygen and silicon with a metal. Rocks are mostly made of silicate minerals.

FIERY ROCKS

Vent

Volcano

Lava flow

Neck

Dike

Sill

Lopolith

Feeder dike

Dike

Batholith

BELIEVE IT OR NOT, a few miles beneath your feet, Earth's interior is super hot—so hot it can actually melt rock! This fiery, molten rock, called magma, is forever pushing up underground. But as it comes up, it cools and turns into new rock, called igneous rock. Magma that erupts from volcanoes also cools to make igneous rocks.

GRANITE

PERIDOTITE

DOLERITE

RHYOLITE

BASALT

WHEN MAGMA WHOOSHES OUT THROUGH VOLCANOES as lava or debris, it cools to form "extrusive" or volcanic igneous rock. When it cools and turns solid underground, it makes "intrusive" igneous rock. Some magma squeezes into spaces to form strange shapes, like dikes, sills, or lopoliths. Deep down, it forms masses called plutons and domes called batholiths. This deep rock is called plutonic.

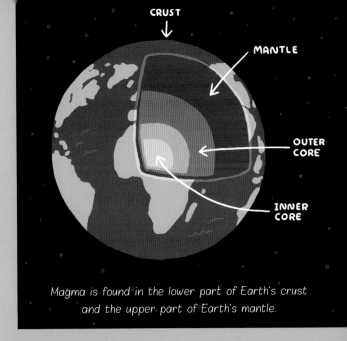

CRUST
MANTLE
OUTER CORE
INNER CORE

Magma is found in the lower part of Earth's crust and the upper part of Earth's mantle.

WHAT KIND OF ROCK?

Deep plutonic rocks have large, coarse grains or crystals, formed when magma cools slowly. Volcanic rocks have tiny, fine grains or crystals, formed when lava cools quickly in the open air. Hot magma that oozes up through the seafloor makes darker rocks than sandy magma that pushes up through Earth's continents.

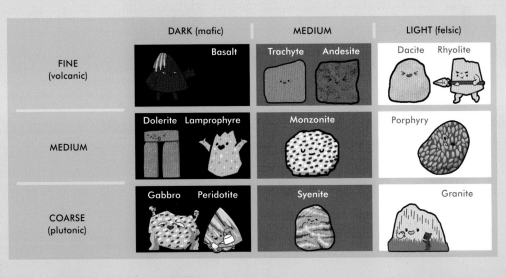

	DARK (mafic)	MEDIUM	LIGHT (felsic)
FINE (volcanic)	Basalt	Trachyte Andesite	Dacite Rhyolite
MEDIUM	Dolerite Lamprophyre	Monzonite	Porphyry
COARSE (plutonic)	Gabbro Peridotite	Syenite	Granite

OBSIDIAN

PERLITE

PUMICE

TUFF

11

GRANITE

STAND-OUT ROCK

DEEP AND TOUGH

WOW!

GRANITE IS BUILT TO LAST! It's one of the toughest of all rocks, and one of the world's hardest natural substances. This speckled rock underpins every continent, and it has done so for many hundreds of millions of years. If you want a tough building material, go for granite. No wonder they used granite for the base of the Statue of Liberty! It's a mineral jumble—coarse-grained with patches of feldspar, a lot of quartz, as well as mica.

QUARTZ MICA FELDSPAR

STANDING PROUDLY

The world's third-highest mountain, Kangchenjunga, at 28,169 feet (8,586 m), is made of granite. The second highest, K2, is made of gneiss, which is even harder. But the world's highest, Everest, is made of softer limestone, so may one day lose its crown!

Granite is a plutonic igneous rock, named for Pluto, the fiery Greek god of the Underworld. Plutonic rocks are forged underground. Long ago, continents smashed together, squeezing rock deep underground, where it melted into giant blobs of magma.

These monster blobs slowly froze into masses called plutons. The biggest of these are batholiths—super-huge lumps of rock, like granite. Most batholiths are still lurking underground, beneath our feet. But in some places, the overlying softer rock has been worn away by weather, leaving tough granite mountains standing proudly like islands in the sea. El Capitan in Yosemite, the Sugarloaf Mountain in Rio, and, of course, Mount Rushmore make unforgettable granite landmarks. Polished, hard-wearing granite is used in everything from fancy buildings to kitchen countertops.

America's first commercial railway, known as the Granite Railway, was opened in 1826 to carry granite for the Bunker Hill monument from the Quincy Quarries in Massachusetts.

Sculptor Gutzon Borglum designed Mount Rushmore.

The presidents were originally going to be sculpted down to their waists!

GRANITE: Grain size: Very coarse • Makeup: Quartz, feldspar, mica, hornblende
Color: Mottled pink, brown, gray with specks • Location: Batholiths, deep underground

PERIDOTITE

THE GREEN GIANT FROM DOWN UNDER

CHROMIUM PLATED

PERIDOTITE IS BIG, VERY BIG! But you don't see much of this green giant because most of it is hidden deep down in Earth's underworld. It's probably the most common rock of Earth's mantle—the main bulk of our planet's interior, which lies between the crust and the core. Millions of years ago, peridotite melted and erupted in very unusual volcanic eruptions, carrying sparkling diamonds with it!

DIAMOND BRINGER

When peridotite oozes up from the underworld, it mostly melts to form basalt (see pp. 20–21). Sometimes, though, chunks of solid peridotite itself end up scraped onto the edge of continents, in a green-brown mess, by the movement of the ocean floor.

In a few, rare places, super-hot peridotite lava from deep under Earth's surface powers up to the surface, forming narrow pipes. These pipes are called kimberlites. They bring up fabulous ancient diamonds, forged deep in Earth billions of years ago. They also bring up rare minerals right from the edge of Earth's core.

Other rocks in peridotite's group, found deep in Earth's interior, are dunite, gabbro, and syenite. Like peridotite, they may be rich in the mineral olivine.

Peridotite brings us one of the world's hardest materials—diamond. It can also alter to make talc, one of Earth's softest materials!

Almost pure olivine, dunite is green like mint ice cream and the source of super-shiny chromium, for car fenders and electric kettles.

Dark, coarse gabbro formed long ago in mushroom-shaped formations called lopoliths.

Hi world!

Sometimes shimmering blue or green, syenite is a lot like granite but has little or no quartz.

Peridotite comes with its own pet gemstone, peridot, made from pure olivine. Peridot is one of the few gems that comes in a single color: deep green.

PERIDOTITE: Grain size: Medium to coarse • Makeup: Olivine, augite • Color: Dull, dark green • Location: Layers in ocean rocks

DOLERITE

THE "SINGING" STONE

NEW YORK PAVER

DOLERITE IS A HARD, DARK-GREENISH rock formed when magma squeezes itself into cracks in existing rock, cools, and becomes solid underground. Where rock beds are in level layers, dolerite magma may intrude in between the layers and freeze into flat, hard platforms called sills. When the softer rock wears away, the dolerite stands out in dramatic cliffs and ledges.

SINGING STONES

About five thousand years ago, British people carried giant slabs of bluestone dolerite hundreds of miles from the Preseli Hills in Wales to make an amazing stone circle: Stonehenge. Why would they go to such a huge effort? It's a big mystery, but experts have a theory. These bluestones are special "singing stones." Instead of giving a dull thud when hit, they make a ringing twang. So maybe they gave Stonehenge music.

The streets of New York were once said to be paved with gold! Not true! But they were once made from chunks of dolerite stone. These chunks were called Belgian blocks. But they didn't come from Belgium; they came from the Palisades Sill, which ranges from New York through New Jersey. The cliffs of dolerite there were given their name in the sixteenth century because they looked like the palisades, or wooden fences, of a fort. So much stone was taken, though, that the authorities banned further extraction. Now most of the blockstone streets are covered in asphalt.

FREEZE!

Sometimes, magma can split rock vertically as it rises. When it freezes, it creates walls of rock called dikes. Some dikes are made from sparkly rock called lamprophyre.

Pegmatites are treasure chests for gems. They are patches of rock packed with big crystals that form from the last dregs of magma. Lithium for phone batteries is also found in pegmatites.

To be a pegmatite, a rock must be made of crystals at least .39 inch (1cm) across.

1cm

In a pegmatite, you might find:

APATITE • GARNET • HELIODOR • BERYL • SPODUMENE • TOPAZ • MORGANITE • EMERALD • TOURMALINE

DOLERITE: Grain size: Medium • Makeup: Feldspar, olivine, pyroxene, mica, others
Color: Dark gray with green tinge • Location: Mostly sills underground

RHYOLITE

FORGED IN EXPLODING VOLCANOES!

MADE FROM MELTED SAND

RHYOLITE IS A SUPER-TOUGH ROCK FORGED BY FIERY VOLCANOES.

When the volcano erupts, it spews magma to the surface, where it's known as lava. Rhyolite lava freezes so quickly that there's no time for crystals to grow, so its grains are tiny. But they are locked so tight that rhyolite is great for paving stones and home exteriors. Prehistoric people hunted animals with rhyolite arrowheads!

ERUPTION!

Rhyolite, andesite, and dacite are all extrusive igneous rocks, meaning that they are made from eruptions of lava. They're all fine or medium grained and pale, because they contain a lot of sand-like silica. They're tough, too. That's why they make good building materials. When you're riding in a car or bus, the road beneath you probably sits on dacite chips.

Rhyolite lava has the highest silica content of all, which it gets when it oozes up through deep layers of Earth's crust. The sand makes the lava so thick and sticky that it may clog up as it bubbles up through a volcano . . . then suddenly burst out in a mighty explosion! That's why rhyolite lava, along with andesite, is linked to the world's most violent volcanoes, like Mount Tambora in Indonesia.

Where there's a cone-shaped volcano, you'll find andesite, with its salt-and-pepper-colored grains. Road-chip dacite, on the other hand, is linked to dome-shaped volcanoes like Mount St. Helens in Washington State.

The ancient Romans made pillars from beautiful reddish volcanic rocks dotted with large crystals. These rare rocks are called porphyries and were highly prized.

Yellowstone National Park is home to one of the planet's largest volcanos.

The last big eruption was about 630,000 years ago but there have been 30 rhyolite lava flows since then!

RHYOLITE: Grain size: Fine • Makeup: Quartz, feldspar, mica • Color: Pinkish-brown • Location: Lava flows, dikes, volcanic plugs

BASALT

HOLDS UP THE OCEANS

"ROAD-MAKER KING"

BASALT IS A DARK, DARK ROCK THAT FORMS FROM THE HOTTEST LAVA OF ALL, reaching nearly 2,000°F (1,090 °C). This burning-hot lava floods up through cracks in the ocean bed, then cools hard when it meets the icy water. Basalt floods spread out to form most of the ocean floor. Chips of basalt are coated in tar to make most of our roads, too.

EARTH'S BEDROCK

Basalt is made from the hottest, darkest lava Earth's interior can deliver. And the shock of cold water or air makes it cool so quickly that it forms tiny grains—so tiny you can barely see them even with a microscope. Though you can try!

Most of Earth's basalt is hidden away under the oceans. But you can see it in places like Hawaii where fountains of basalt lava burn through the crust to form volcanoes such as Mauna Loa. You can see it in India, too, where at the time the dinosaurs died, vast floods spilled out to make what we now call the Deccan Traps. From 17 to 14 million years ago, similar floods poured out to form the vast Columbia Plateau in the northwest United States. You can see basalt on the moon, too, where it forms dark patches called "seas."

On the coast of Northern Ireland, you can see strange-looking rocks in hexagonal columns. Locals of old said they must be made by giants, and called the rocks the Giant's Causeway. But they're just basalt cracking this way as it cools and hardens.

Mars has the solar system's biggest volcano. It's called Olympus Mons, and it's about the size of Arizona. It's made largely of basalt.

The moon is made of CHEESE!

Actually it's not. About 26% of the near side is basalt "seas."

BASALT: Grain size: Very fine • Makeup: Feldspar, pyroxene, olivine, magnetite • Color: Black • Location: Lava flows, dikes, sills

OBSIDIAN

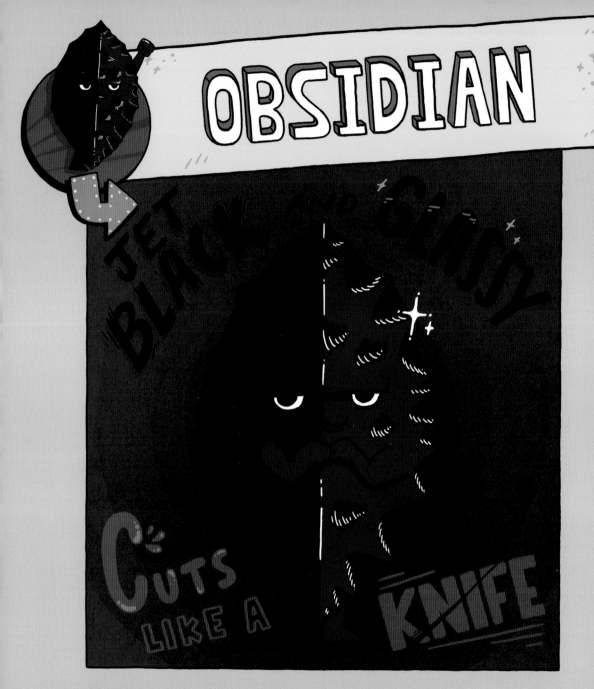

JET BLACK AND GLASSY

CUTS LIKE A KNIFE

OBSIDIAN IS LIKE NO OTHER ROCK. Its lava solidifies so suddenly that no crystals can form, so it's like shiny black glass. It breaks like glass, too, to give a super-sharp edge. Back in the Stone Age, obsidian gave the sharpest knives imaginable. Night-black obsidian knives looked amazing, too. No wonder this rock was sacred and mysterious for everyone from the ancient Egyptians to the Mayans.

OBSIDIAN: Grain size: None • Makeup: Quartz, feldspar, and mica • Color: Black • Location: Lava flows, and dikes

PERLITE

Perlite is made from the same lava as obsidian, but it's the color of dirty snow. Bubbles of water trapped in the lava as it freezes turn to steam, and blow up the rock like popcorn—to twenty times its original size. This creates a super-light rock, full of gas bubbles, that's useful for all kinds of jobs—from insulating pipes to making lightweight concrete. Gardeners add it to soil to make it dry and airy, good for growing plants.

Perlite: Grain size: None • Make-up: Quartz, feldspar, mica
Color: Light gray • Location: Lava flows, dikes

Pumice is the only rock that floats! When rhyolite lava erupts, the top fizzes up with air bubbles—like opening a shaken can of soda pop. This froth blows the lava into the air. Where it falls, it hardens into superlight chunks. Pumice's rough surface was once used to scrub skin clean. Now pumice is used to "stonewash" jeans, giving them a fashionable faded look.

Pumice: Grain size: None • Make-up: Quartz, feldspar, mica
Color: White • Location: Lava flows, fragments

PUMICE

TUFF

When a volcano explodes, some of the solid rock blocking its pipe gets smashed into ash and blasted high into the air. When this ash finally falls back to the ground, it lies in a thick blanket that eventually hardens into a light, soft rock called tuff. In 79 CE, Mount Vesuvius erupted in Italy, covering the Roman city of Pompeii in ash. The ash turned to tuff, preserving the city's tragic last day for centuries.

Tuff: Grain size: None • Make-up: Quartz, feldspar, mica
Color: Buff, gray • Location: Ashfalls

PILED-UP ROCKS

Rocks are broken into fragments by the weather. The fragments are carried by rivers into lakes and oceans.

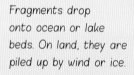

Fragments drop onto ocean or lake beds. On land, they are piled up by wind or ice.

More fragments drop on top, and all the squeezing drives out air and water.

SHALE

SANDSTONE

CHALK

CHERT

FLINT

IGNEOUS ROCK IS SUPER TOUGH, but after being exposed to the weather for a long time, it crumbles to bits! The bits become pebbles, sand, and gravel, which are washed down rivers to the ocean to settle on the ocean floor, or piled up by wind in deserts, as "sediments." Sediments pile up in layers, one on top of the other. Over hundreds of millions of years, the layers are squeezed dry, baked hard by Earth's hot interior, and glued together by minerals to form sedimentary rock.

Not all sedimentary rock is made like this. Some, like travertine, is made from fine powder left by dissolved minerals. Others, like limestone, are made from the remains of living things.

The shifting of Earth's surface means that sedimentary layers or "beds" are rarely left in peace. Some are thrown up as new hills and mountains. You can see lots of beds exposed like a layer cake in the Grand Canyon, where the Colorado River has cut down through them. The newest rocks are at the top. The oldest rocks are at the bottom.

The rock type varies from layer to layer, because each was deposited in particular times and conditions. The layers are called rock beds or strata.

The fragments are glued together by a paste of minerals, turning them into hard layers of rock.

LIMESTONE   TRAVERTINE   DRIPSTONE   TUFA   COAL

SHALE

SHALE IS LAYERED, LIKE THE PAGES OF AN ANCIENT BOOK. It's made from mud that settles on ocean floors and lake beds. The mud is squeezed hard, then dries out over millions of years and cracks into layers. Shale may look a little dull, but it's the universal builder—bricks, cement, and pottery all contain shale. And the oil and gas that we rely on every day are both found in shale, too.

MUDROCKS

Shale is one of various kinds of mudrocks. You can see mudrocks everywhere—they're the most common sedimentary rocks of all. And yes, you guessed it—they all started as mud. Mud is basically a soggy mush of tiny grains called silt or even tinier grains called clay. Clay is the last, tiniest bits left when rocks are broken down by the weather.

Clay grains are so light, they are often washed far out into the sea or into a lake before settling on the sea or lake bed. As they settle, they become mud. In time, more clay settles on top, squeezing the mud to make claystone, one of the softest of all rocks. Shale forms when claystone gets squeezed even harder and becomes even drier.

The "clay" potters use to make pots is actually clay-rich mud or claystone. To geologists, it's like oatmeal. The oatmeal is the mud, and the oat grains are the clay.

People have molded soft clay into pots and plates since ancient times. Those made from red-brown clay look like flowerpots. Certain clays, such as kaolinite, are treated with heat to make beautiful white porcelain.

Clays can be good for preserving fossils. Some of the first fossil hunters, including Mary Anning (1799-1847), discovered amazing prehistoric marine reptiles, called plesiosaurs, in the Oxford and London Clays of England.

I'll huff and I'll puff and I'll blow your house in!

Good luck! The bricks are made from clay!

SHALE: Grain size: Fine • Makeup: Quartz, feldspar, olivine, mica • Color: Black, gray, dark brown • Location: Sedimentary layers

SANDSTONE

TOUGH AND GRITTY

RIPPLED & LAYERED

SANDSTONE IS JUST WHAT IT SAYS IT IS—ROCK MADE OF SAND! The sand comes from other rocks that have crumbled, and is carried by wind, water, or ice to river deltas, beaches, or deserts. Then it's squeezed and squeezed, and cemented, or glued, into new rock by minerals. Beach sand is light tan, but the color of sandstone depends on its mineral cement. Limonite cement gives yellow sandstone. Iron oxides are rusty to give red or brown sandstone.

STURDY SURVIVORS

Sometimes, the cement in sandstone is so frail that the rock crumbles like cake. But mostly sandstone is super tough! It defies the wind and rain and weather to create some of the world's most dramatic landscapes. The buttes (tower rocks) and mesas (fortress-like plateaus) of the deserts of Utah and Arizona are gritty survivors that stood firm while the softer rocks around them were worn away.

Sandstone is tough, yes, but when still fresh and damp from the ground, it's quite easy to cut. That makes it a perfect building material. In the 1800s, tens of thousands of houses in New York were built with brownstone, a brown sandstone dug out of the ground at the famous Portland Quarries in Connecticut.

Conglomerate is made of pebbles rather than sand. Random pebbles are washed up by rivers, then cemented together. Breccia is like conglomerate with sharp chunks.

The Old Red Sandstone are some of the world's largest sandstones. They formed from sand that piled up in a ginormous desert basin in what is now northwest Europe, 360 to 408 million years ago.

The White House was made from gray-white sandstone.

In 1818, they decided to paint it white.

SANDSTONE: Grain size: Sand-sized • Makeup: Quartz, feldspar • Color: Yellow, brown, red • Location: Sedimentary layers

CHALK

CHALK IS SO MILKY WHITE, SO SOFT AND POWDERY-FINE, YOU CAN WRITE WITH IT ON A BLACKBOARD. It's made from the crushed remains of algae and ocean-dwelling microorganisms called foraminifera. They lived in subtropical seas about a hundred million years ago in the Cretaceous period (when the dinosaurs lived on land). When they died, they floated down to the ocean floor and slowly hardened into chalk. Sometimes you can see tiny, tiny fossils in chalk, like a window into the past.

CHALK: Grain size: Fine like mudstone • Makeup: Calcite • Color: White • Location: Sedimentary layers

CHERT IS A TOUGH, GLITTERY, TYPICALLY BROWNISH, GLASSY ROCK MADE FROM TINY QUARTZ CRYSTALS.

It's solidified ooze, the slime from the ocean floor, often made from the remains of animals that turn slowly to silica (the sandy material that makes quartz). Although it is tough, when you hit it with a hammer, it shatters like glass into tiny shell-like shapes. Its toughness, and the fact that it can be chipped to give a sharp edge, made chert one of the perfect stones for Stone Age axes.

CHERT: Grain size: Too small to see
Makeup: Mainly quartz
Color: White to black
Location: Sedimentary layers

FLINT NODULES LOOK LIKE REALLY HARD, KNOBBY LUMPS OF TOFFEE.

Like chert, flint is pure silica, but it's made from chemicals in limestone rather than plankton. Flint nodules can be broken to give an edge as sharp as glass. Obsidian is pretty rare, so when Stone Age people wanted a knife edge, flint was the go-to material. Stone Age people also made fires by hitting flint against another rock to make a spark. Old-fashioned guns called flintlock pistols used flints to create the spark to light the gunpowder to fire, too.

FLINT: Grain size: Too small to see
Makeup: Mainly quartz
Color: Black
Location: Sedimentary layers

LIMESTONE

BLOCK ROCK

THE ROCKS, OF LIFE

BILLIONS OF CREATURES DIED TO MAKE LIMESTONE! There are vast, thick beds of limestone all over the world. They form when calcite and aragonite become solid in calcium-rich water. Mostly, the minerals came from the crushed remains of corals and shellfish and other sea creatures that lived in the ocean hundreds of millions of years ago. Look closely, and you can often see how limestone is packed with their fossilized remains.

FOSSIL ROCK

Limestone is the third most common sedimentary rock, after mudstone and sandstone. Most of the limestone we see is ancient, though it's still being made today.

After limestone forms, it dries and cracks into giant blocks. It's a tough rock, but it's made of the mineral calcite, and calcite dissolves in rainwater. So when rain seeps into the cracks between the blocks, it dissolves the rock. Amazing holes and gorges are eaten out of the ground, and giant underground caverns open up. This kind of incredible limestone scenery is called karst.

Despite this, limestone is so tough that it makes a great building material, either as stone, or crushed into a powder to make cement and concrete. Without limestone, very few of our modern cities could be built! It makes roads and bridges, skyscrapers and airports.

Dolostone is like a sugar lump! It's a super-hard, sugary-white form of limestone that formed in tropical lagoons in ancient times. It's named after the Dolomite mountains in Italy, which in turn were named after French geologist Déodat de Dolomieu, who identified the rock they are made from. It's full of magnesia, which is good for cattle feed and in glass-making.

Florida is built on limestone, which dissolves fairly easily in rain water.

Sinkholes can form without **warning!**

LIMESTONE: Grain size: Varies • Makeup: Calcite, aragonite • Color: White, gray • Location: Sedimentary layers

TRAVERTINE

TRAVERTINE IS A HARD, WHITE, POWDERY CHEMICAL CRUST, LEFT BY HOT MINERAL WATER WHEN IT EVAPORATES. The famous white terraces around the hot springs in Yellowstone National Park are travertine. At Mammoth Hot Springs, an entire hill of travertine has built up! Travertine's not quite as beautiful and sparkly as marble, but sculptors still love it. In St. Peter's Square in the Vatican, where the Pope lives in Rome, the beautiful columns are made from travertine.

TRAVERTINE: Grain size: Fine • Makeup: Calcite • Color: White • Location: Caves, hot springs

DRIPSTONE

DRIPSTONE IS STALACTITE AND STALAGMITE ROCK. STALACTITES LOOK LIKE ICICLES; they form from the minerals left behind by water dripping from cave roofs. Stalagmites are like fingers building up where water drips onto cave floors. Together, they can make caves look like cathedrals, with narrow columns and organ pipes! Slice across dripstone, and you can see rings showing how it built up in layers. The longest stalactite in the world is 92 feet (28 m) long, in the Gruta do Janelão, in Minas Gerais, Brazil. The tallest-known stalagmite is over 230 feet (70 m) in height and is in Sơn Đoòng Cave, in Vietnam.

DRIPSTONE: Grain size: Fine
Makeup: Calcite
Color: Honey, red, brown
Location: Caves

TUFA

TUFA IS CREAM COLORED AND FULL OF HOLES LIKE A SPONGE, WHICH MAKES IT INCREDIBLY LIGHT. Like dripstone, it forms from the minerals left behind by evaporating water. The holes are formed by algae that later rotted away. The ancient Romans loved it for building intricate structures like roads and aqueducts. The holes make it easy to cut, and the lightness makes it easy to carry. Today, tufa is used in backyard features.

TUFA: Grain size: Fine
Makeup: Calcite
Color: White, yellow, red
Location: Streams, springs

COAL

THE ROCK THAT BURNS

DARK FOSSIL ROCK

SHINY BLACK COAL IS THE FIRE ROCK, THE ONLY ROCK THAT BURNS.
It's made from the crushed remains of plants and trees that grew millions of years ago.
When they died, they rotted and turned blacker and blacker as they were squeezed by
more forests dying on top of them. The blackest coal is hard and shiny, and there
is no other rock that looks anything like it.

FOSSIL FUELS

Much of the coal in North America, Europe, and northern Asia started in Carboniferous times, some 300 million years ago. At this time, these areas were tropical, and there were vast, steamy swamps, thick with plants. When plants died, they piled up in thick layers. Tiny microorganisms rotted these layers into dark brown, earthlike peat. As the peat was buried by more sediments, it was squeezed hard and turned into thin layers, or seams, of black coal. The deepest buried coal is shiny black and almost pure carbon.

Coal is a "fossil fuel"—a natural fuel formed from the remains of living things. The carbon burns in air to create heat energy. Coal was used to make fires for centuries, but in modern times, its energy drives the turbines that make electricity.

The deeper coal is buried and the more it's squeezed, the hotter and cleaner it burns. The least squeezed and smokiest is peat. Then comes brown coal, coal, and then very hot and clean anthracite.

To get the best deep coal, mining companies sink a shaft into the ground, then extend tunnels out to get at the beds, or seams, that hold the coal. It's dangerous and difficult work.

The Bagger 293 digs brown coal near Hambach in Germany.

It's the world's biggest vehicle. It's 738 feet (225m) long and 315 feet (96m) tall.

That's about as wide and a bit taller than the U.S. Capitol Building!

COAL: Grain: None • Makeup: Carbon • Color: Black • Location: In between sedimentary layers

CHANGED ROCKS

ROCKS MAY BE TOUGH, BUT WHEN THEY ARE SCORCHED by hot magma or crushed by the mega pressures of the shifting of Earth's crust, they can be altered beyond recognition. Heat and pressure changes the mineral content—grains and crystals—so completely that they become new rocks, called metamorphic rocks.

HORNFELS

MARBLE

GNEISS

SLATE

Contact metamorphism: Wherever magma rises, it can cook the rock above it, like a cake. This is called contact metamorphism, and the effect can be truly dramatic!

QUARTZITE

MARBLE

HORNFELS

SANDSTONE

PURE LIMESTONE

SHALE AND CLAY

Regional metamorphism: The mega pressure created by the shifting of Earth's crust, especially when mountains are built, squeezes rock hard over vast areas. This is called regional metamorphism. It develops in "grades" as the pressure rises.

LOW GRADE
Clay and shale ➔ slate:
The rock breaks into sheets.

MEDIUM GRADE
Slate ➔ schist:
The rock becomes stripy as minerals separate into layers.

HIGH GRADE
Schist ➔ gneiss:
The stripes become swirling bands.

With each grade, new rock forms, becoming harder, more glittery, and more layered.

HORNFELS

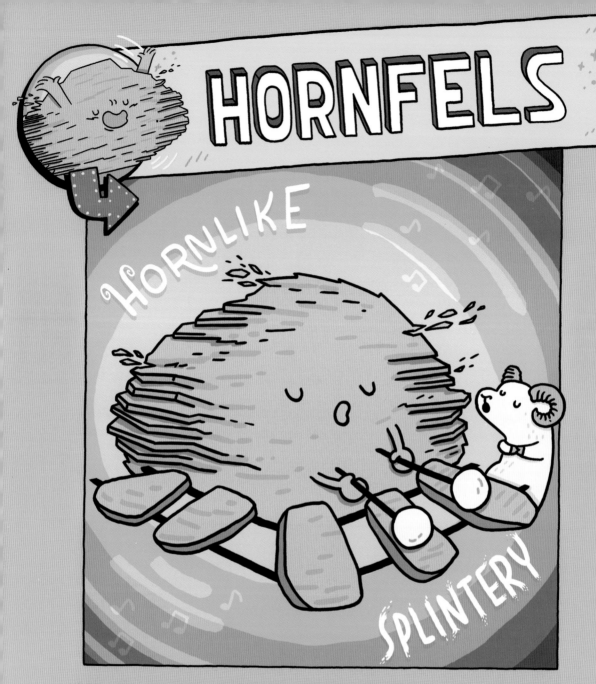

HORNLIKE

SPLINTERY

HARD AND SPLINTERY, THIS ROCK IS CALLED HORNFELS because broken pieces of it look like rams' horns! Hornfels is also known as "ring stone" because it rings when it's hit. People have even made musical instruments out of hornfels! Look under a microscope to see that hornfels' small, equally sized grains interlock snugly, like pieces of a mosaic.

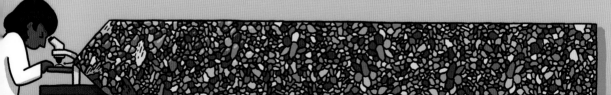

IN CONTACT

When scorching magma oozes up through the rocks of Earth's crust, it's so hot that it literally cooks the rocks around it. As they cook, the rocks almost melt and turn very soft. Their crystals entirely lose their shape. As they cool and harden again, they reform as interlocking fine grains, pointing this way and that, and sometimes even grow new minerals.

Because there is no real pressure—just the heat of the magma—this is called contact metamorphism. Hornfels is not actually a single rock. It is the term for a whole assortment of rocks that form in this way. The end result depends on the heat, but also on the original, uncooked rock, which can be shales and clays, limestone, and volcanic rocks such as basalt.

Back in the nineteenth century, lots of people played Rock Harmonicons, hitting different-sized chunks of hornfels rock. The famous Till family played on Skiddaw stone hornfels from England's Lake District.

Near Hagi, Japan, is a beautiful, banded stretch of coast called the Susa Hornfels. The cliffs are a Japanese national monument.

HOW TO COOK A ROCK

① Take one lump of cold clay (or limestone if you haven't got clay).

② Pour hot, molten magma on it.

③ Leave to cook for several years at 1,350°F (732°C).

④ And BINGO! New hornfels rock!

HORNFELS: Grain size: Fine • Makeup: Mica and many others • Color: Gray, black, brown, often speckled
Location: Next to volcanic intrusions

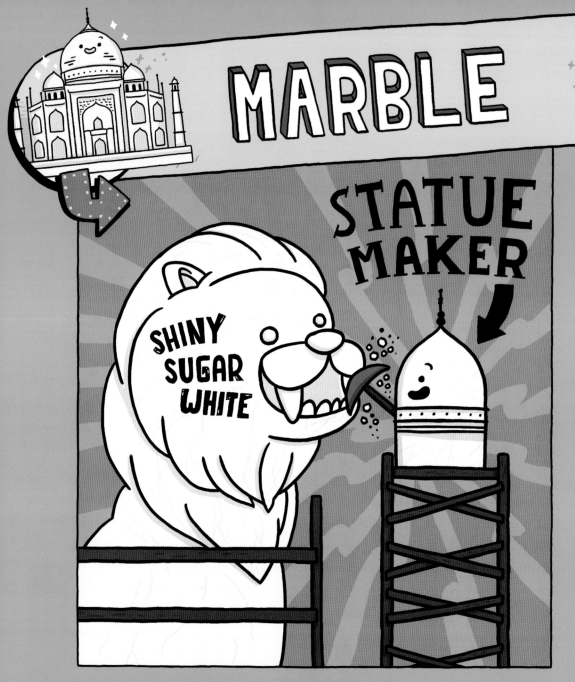

MARBLE

STATUE MAKER

SHINY SUGAR WHITE

MARBLE IS THE SHINIEST ROCK OF ALL. It's like polished sugar, white when it's pure, and even better with beautiful streaks of color. It's soft enough to carve easily—a sculptor's dream stone. It even seems to glow softly because the crystals let a little light through.

SHINING STONE

All marble starts off as limestone. Like limestone, marble is made mostly of calcite. But while limestone is gray and dull, marble is calcite purified by extreme heat and pressure. The limestone that makes marble is totally crushed by the weight of mountains above and cooked seriously hot by the heat of Earth's interior.

We probably don't see most of the marble made underground. Just a little is brought up from the mountain roots by the shifting of the continents. Then you can see gray limestone transformed into shiny marble. It's mostly white but impurities in it are smudged out into wonderful, whirling, colored ripples. Sometimes the whole marble is stained. Pyroxene stains it green; sphene, yellow. Graphite turns it a beautiful black.

The town of Marble in Colorado gets its name from the Yule Marble Quarry, which operated from 1905 until 1941. It was the source of the marble that was used for the exterior of the Lincoln Memorial in Washington, DC.

The world-famous Taj Mahal in India was built between 1631 and 1648. It looks like a fairy-tale palace or a temple. But it was built on the orders of the emperor, Shah Jahan, to house the tomb of his wife Mumtaz Mahal.

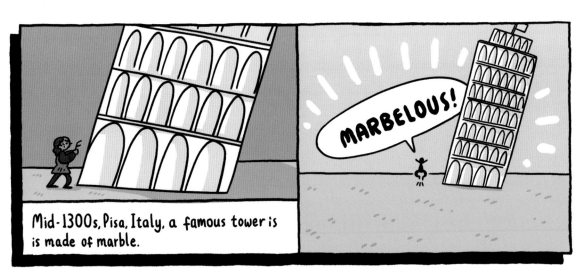

Mid-1300s, Pisa, Italy, a famous tower is is made of marble.

MARBELOUS!

MARBLE: Grain size: Medium to coarse • Makeup: Mainly calcite • Color: Sometimes white, can be stained or rippled
Location: Limestone mountains

GNEISS

SUPER OLD

SUPER TOUGH AND SUPER SPARKLY

SPARKLY GNEISS IS THE OLDSTER OF ALL THE ROCKS ON EARTH. It's so tough—it's the toughest rock of all—that almost nothing damages it, and it just never, ever wears away. It's typically gray or pink with dark black tiger stripes. Some gneisses have beautiful red garnets embedded in them.

GRANDPA ROCK

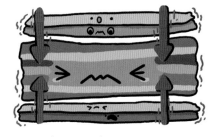

Gneiss is forged under huge pressure. Earth's surface is made of a jigsaw puzzle of twenty or so giant, slowly-moving chunks called tectonic plates. Gneiss forms where two of these plates crush together with unimaginable force. The rock is crushed so hard that it dissolves and reforms. It doesn't matter what the original rock is—it's squeezed to make gneiss.

Gneiss is stripy, with dark and light bands of minerals smeared out like stripy toothpaste, making gneiss tough. The light bands are usually feldspars and quartz; the most important dark minerals are hornblende and biotite. Vast masses of ancient rocks are made from gneisses that formed long ago. Large areas of Greenland are made of gneisses that are at least three billion years old, and some of the world's oldest rocks are Acasta gneisses of northern Canada, over four billion years old!

Schist (see p. 39) is made when shale or mudstone are squeezed really hard. Like slate, it's made of sheets, but the sheets are super thin and fused together.

The Washington Monument in Washington DC, completed in 1884, is the world's tallest stone monument, at 555 feet 5 1/8 inches high. It's built of marble and granite but relies on a core of super-tough bluestone gneiss, dug from quarries in the Potomac valley.

People have used garnets to wish others PEACE and HEALTH ...and friendship!

GNEISS: Grain size: Medium to coarse • Makeup: Typically feldspar, quartz, mica
Color: White, red, pink, brown, gray, black • Location: Very deep

SLATE

BLACKBOARD GRAY

SMOOTH AND FLAKY

IMAGINE SOME MUDSTONE (A TYPE OF MUDROCK, SEE P. 26) OR SHALE. Now imagine squeezing it really hard. Now imagine it getting squeezed underground unbelievably, incredibly hard. Now it's slate, a dark gray, tough, slightly shiny rock. All the clay grains inside it got flattened and regrew in thin layers. So slate splits easily into thin sheets at right angles to the pressure.

KEEPS US DRY

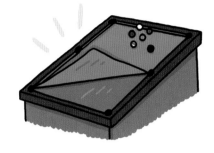

For thousands of years, skilled slate workers have been digging out slate and hitting it. When hit just right, it splits into smooth, flat sheets of stone. These slate tiles are perfect for roofs, because they are light, tough, and almost completely waterproof.

When cities grew during the Industrial Revolution of the late 1800s, there were slate booms, when everyone wanted to put slate roofs on their new homes. More than half a million tons of slate a year were coming out of the slate quarries in Wales in the United Kingdom. Across the Atlantic, major amounts were mined in quarries in the United States, in Vermont and in Pennsylvania. Today, most roofing slates are not natural slate at all but made from fiber and cement or concrete.

Under the green felt of the best pool tables, there's always slate. Slate can be ground and polished to give a smooth surface that's perfect for pool.

In the old days before computer screens, teachers often relied on two rocks: chalk and slate. The slate was used to make a blackboard, which the teacher wrote on in white chalk.

Slate Quarry

Blow the charges!

Let's get cutting!

Let's split!

SLATE: Grain size: Very fine • Makeup: Clay minerals, mica • Color: Black, gray, dark brown
Location: Formed during mountain building

MAGIC MINERALS

MINERALS ARE THE NATURAL MATERIALS ROCKS ARE MADE OF.

Wherever there are rocks, there are minerals! They're nearly all made of solid crystals. Some are too small to see, but if you know where to look, you can sometimes find big, beautiful crystals, or gems. It's these that mineral collectors usually want. So, on the opposite page, you can see some of the situations where they occur.

Some mineral superstars:

BARITE

GYPSUM

CROCOITE

WULFENITE

MONAZITE

HOT SPOTS

Here are some situations in which great crystals form.

Vein: When hot liquid from magma oozes into cracks deep inside Earth, the fluid often cools to leave fabulous gems such as calcite, feldspars, galena, gold, rhodochrosite, topaz, and tourmaline.

Pegmatite: Pegmatites (see p. 17) form from the last super-concentrated bit of magma, which cools slowly, forming treasures such as apatite, beryl, garnet, aquamarine, tourmaline, topaz, fluorite, and corundum.

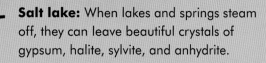

Sediment pocket: Old pockets in rocks, once filled with fluids, can make little treasure chests, too. Here, look for barite, calcite, chalcedony, sphalerite, pyrite, and turquoise.

Salt lake: When lakes and springs steam off, they can leave beautiful crystals of gypsum, halite, sylvite, and anhydrite.

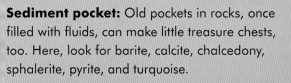

HALITE TURQUOISE FLUORITE RUBY SPINEL

MINERAL CHEMISTRY

Gold

There are thousands of kinds of minerals, each with its own special chemistry. It all depends on their mix of elements—the 100 or so basic chemicals from which everything is made. A few minerals, such as gold, are made from a single element. But most belong to families or compounds. Compounds are partnerships of two or more elements.

Native elements:
These are mostly found in igneous and metamorphic rock, but some toughies, such as gold and diamond, survive after the rock is broken to be washed into streams. *Gold, copper, silver, diamond, sulfur*

Sulfides: These typically form in veins with hot, rising water or from magma. They are often brittle and have a metallic look. *Sulfur and a metal, such as lead*

Lead

Sulfur

Primary oxides: These are hard, like corundum, and form deep down in magma or in hot veins. Soft, secondary oxides such as bauxite form when minerals such as sulfides are attacked by the air. *Oxygen and any metal except gold or silver*

Halides: These are salts like table salt. Except for common salt, they are rare because they dissolve in water. *Metals with a halogen element such as chlorine and bromine*

Carbonates: These are brought to the surface in hot fluids, but most form when surface minerals, like malachite and azurite, are altered. *Carbon and oxygen with metals or semimetals*

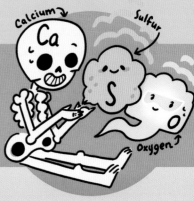

Sulfates: These are mostly soft and pale, like gypsum and barite. *Sulfur and oxygen with metals*

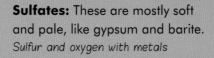

Phosphates: These are rare but often have vivid colors, such as turquoise. *Phosphorus and oxygen with metals*

Silicates: There are more silicates minerals than all the other minerals put together. Most rocks are built with a LOT of the silicates like quartz and feldspars. *Silicon and oxygen, often with metals*

REALGAR AND ORPIMENT

THESE MINERAL TWINS LOOK SO PRETTY! BUT BEWARE! Realgar is a brilliant cherry red. Orpiment is yellowy-orange, like butter. That's why painters once used them as pigments. They were used for fireworks, too. But don't be deceived! Hit them with a hammer and they give off a bad whiff of garlic. And they contain super-poisonous arsenic. So don't touch them! Ever!

REALGAR

Realgar is made in steamy volcanic waters from sulfur and dangerous arsenic. It's so poisonous that in ancient times people used it to deter rats, flies, and snakes. It's actually still used (with great care) as an ore for arsenic, which is useful for everything from medicines to electronics.

You could call realgar the vampire mineral! Not only is it blood-red, but it crumbles to dust when exposed to sunlight. That's why geologists keep samples in the dark. Its name comes from the Arabic rahj al-ghar, which means "powder of the mine."

REALGAR: Group: Sulfides
Makeup: Arsenic sulfide • Color: Orangey-red
Streak: Orange • Luster: Resinous, pearly
Hardness: 1.5–2 • SG: 3.5

From ancient times, arsenic was used to kill, so poisonous that it could be powdered into your victim's food or drink in such small amounts they'd never notice. The Borgias, a powerful Italian family in the fifteenth and sixteenth centuries, were said to have killed many of their rivals with arsenic!

On New Zealand's North Island, there's a volcanic spring called Champagne Pool. The water has traces of gold, silver, mercury, arsenic, and antimony. The bubbles are carbon dioxide. The rim is bright orange, created by orpiment and realgar.

ORPIMENT

Since ancient times, people have used orpiment as a pigment to add a yellowy-orange color to all kinds of things, without realizing just how dangerous it is.

You can find it in the tomb of Tutankhamun, the boy pharaoh of Egypt, who died in around 1323 BCE. It was also treasured in ancient China, where they used it to make silk a bright yellow. Painters used it to make a yellow color until 200 years ago, when they began to make yellows from the element cadmium, instead. Nobody knows just how many people were made ill or even killed by orpiment over the centuries.

ORPIMENT: Group: Sulfides • Makeup: Arsenic sulfide
Color: Orangey-yellow • Streak: Yellow
Luster: Resinous, pearly • Hardness: 1.5–2 • SG: 3.5

GALENA

GALENA FORMS SHINY, SILVERY CRYSTALS THAT LOOK LIKE SOMEONE HAS MADE METAL BRICKS, THEN TUMBLED THEM IN A PILE! Ancient people thought that galena was magical. Crushed galena was the earliest eyeliner, worn by pharaohs and noblepeople across the ancient world. If you melt galena down, it's full of lead. Lead is soft, gray, and super heavy, and it's a useful metal. The ancient Romans used lead for making pipes and to line their aqueducts, because it doesn't corrode. They didn't realize that lead is poisonous.

GALENA: Group: Sulfides • Makeup: Lead sulfide • Color: Dark gray • Streak: Lead gray
Luster: Metallic to dull • Hardness: 2.5+ • SG: 7.5–7.6

PYRITE

SO MANY PEOPLE HAVE MISTAKEN CHUNKS OF PYRITE FOR CHUNKS OF GOLD THAT IT'S SOMETIMES CALLED "FOOL'S GOLD."

Though pyrite's made from iron and sulfur, it's not at all dull. When struck hard against other rocks, it makes a bright spark. That's why it was used to light fires in ancient times, and its name comes from the ancient Greek word for "fire." It's very common, and chances are that any rock that looks a little rusty has got a little pyrite in it, because the iron it contains goes rusty in the air.

PYRITE: Group: Sulfides • Makeup: Iron sulfide
Color: Brassy yellow • Streak: Greenish black
Luster: Metallic • Hardness: 6–6.5 • SG: 5.1+

CINNABAR

CINNABAR IS BRIGHT RED—ITS NAME COMES FROM THE ANCIENT PERSIAN FOR "DRAGON'S BLOOD."

But watch out! It's made from poisonous mercury. In the past, people hunted for it near the volcanoes and hot springs where it forms, using it to make paint colors such as vermilion. Some women even used it for cosmetics, but sadly, they didn't know it was slowly poisoning them. Fortunately, the cinnabar you see in shops today is safe, "imitation" cinnabar, but take care until you're sure.

CINNABAR: Group: Sulfides • Makeup: Mercury sulfide
Color: All shades of red • Streak: Red
Luster: Adamantine • Hardness: 2–2.5
SG: 8.0–8.1

AWESOME ORES

OUR WORLD IS BUILT ON METAL.

There's metal in everything from cars and houses to smartphones and saucepans. It all comes from rocks. Rocks are full of metals, but they're not easy to see. If you're super lucky, you might find a few lumps of gold or copper in the ground. But most metals are locked away inside special minerals in rocks called ores.

There are thousands of different minerals, but only about 100 are ores.

Galena is rich in lead, which is why it's our main lead ore.

Iron is found in several ores, such as hematite and magnetite.

Pyrite contains a lot of iron, too, but it's so hard to get out that no one bothers!

NOT LEAVING!

Some ores contain more than one metal. So if you extract lead from galena, you might get silver as a bonus!

HOW DO YOU GET METAL OUT?

- Dig out the rock, quarry it from the surface, or mine it from deep underground.
- Separate the ore from the rock by crushing, washing, and flltering, or even by using magnets and electricity.
- Get the metal from the ore by "smelting," heating until the metal melts and runs out. You might have to use electricity or chemicals instead of heating.
- Refine or purify the metal or mix it with other metals to make an "alloy."

Neodymium

We need "rare earth" metals, such as neodymium, dysprosium, and gadolinium, for modern tech such as smartphones and electric car batteries. They're difficult to extract! The main mineral sources are monazite and bastnaesite.

Introducing! ORES SCIENCE FAIR

Here are some of the main ore minerals and the metals they contain.

Bauxite presents . . .

Aluminum

Hematite, magnetite, and siderite present . . .

Iron

Halite presents . . .

Sodium

Cuprite and malachite present . . .

Copper

BARITE

SOLID PERFORMER

THE HEAVYWEIGHT

WHITE OR COLORLESS, BARITE IS A HEAVYWEIGHT. Also known as heavy spar, its name comes from the Greek baryos meaning "heavy." Ground barite mud is pumped into oil drills to contain the oil. Barite is so dense that it's also made into bricks to seal off dangerous radiation in hospital X-rays and scanners.

THE SMOOTHIE

Barite doesn't seem like the most exciting mineral. It rarely forms nice crystals, and it's usually a dull, dirty white. But barium makes a very smooth, heavy powder. This is added to playing cards to give them a silky finish for easy dealing!

Sometimes barite can solidify into perfect rose-like shapes, formed from groundwater seeping through sandstone. Most barite roses range from .4 inches to 4 inches (1 cm to 10 cm) in diameter, though the largest one is five times that. Each "petal" is a single crystal of barite. The barite rose became the official state rock of Oklahoma in 1968.

Barium's cousin, celestine, is a beautiful, pale blue mineral. It forms in sediments under the sea as water trickles through them. It's a source of strontium, which is used for fireworks and burns red, not blue..

If you have a gut problem, the doctor may ask you to take a drink containing barium powder, called a barium meal. The barium meal is so dense that it shows up on X-rays, revealing the shape of your gut.

Watch Out! The oil could shoot out!

I added heavy barite mud to the drill.

The weight will stop the blowout.

Thank barite!

BARITE: Group: Sulfate • Make-up: Barium sulfate • Color: Colorless or white • Streak: White
Luster: Glass • Hardness: 3–3.5 • SG: 4.5

GYPSUM

KEEPING IT SMOOTH

PLASTER AND FILLER

GYPSUM IS A VERY COMMON MINERAL, AND IT'S SUPER USEFUL.

If you're indoors, go to your walls and touch them. The chances are high that you're touching gypsum! Gypsum is the powdery mineral that's smeared on your walls as plaster to make them smooth.

CRYSTAL LEFTOVERS

Whenever salty water evaporates, it leaves salts behind, and gypsum is one of those salts. That's why thick beds of soft gypsum rock form all over the world in places where there were once shallow seas or lagoons. Gypsum rock crumbles easily to a fine soft powder, and that's what the plaster for our walls is made of.

You'll find gypsum in other rocks, too, such as limestone and sandstone. And it's not always a fine powder. Sometimes it forms in long, thin, silky white crystals called satin spar. Or it can occur in chunkier milky-colored crystals called selenite. Satin spar and selenite look very different from gypsum rock, but they are all basically gypsum. We eat possibly 28 pounds (12.7 kg) of gypsum in our lives because it's used to bulk up food like ice cream, spaghetti, and bread—even tofu! It's harmless when processed in these foods, but don't ever eat it otherwise.

Gypsum rock sometimes forms as a beautiful, pale stone called alabaster. Sculptors have made statues from it since ancient times, because it's so smooth and easy to carve.

Covering a huge area of 275 square miles (710 sq km) in New Mexico, White Sands National Park is full of spectacular dunes of pure white gypsum.

HOW DO YOU TURN GYPSUM ROCK INTO PLASTER FOR YOUR WALLS?

First you **CRUSH** it

Then you **GRIND** it

Then you **BAKE** it in a kiln...

GYPSUM: Group: Sulfate • Makeup: Calcium sulfate • Color: Colorless or white • Streak: White • Luster: Glassy
Hardness: 2 • SG: 2.3+

CROCOITE

CROCOITE GETS ITS INTENSE ORANGE-RED COLOR FROM THE PRESENCE OF THE ELEMENT CHROMIUM. It grows crystals in dramatic needles, looking like a crazy orange pincushion. Giant crocoite crystals, up to four inches (10 cm) in length, are found in Dundas, Tasmania (which is part of Australia), but most are much smaller and very fragile. Crocoite gets its name from being the same color as a kind of crocus that has saffron in its pollen, the super-expensive orange-red spice.

CROCOITE: Group: Chromate • Makeup: Lead chromate • Color: Orangey-red
Streak: Orangey-yellow • Luster: Adamantine or glassy • Hardness: 2.5–3 • SG: 6.0+

WULFENITE FORMS THE MOST AMAZING, THIN ORANGE-YELLOW INTERLOCKING PLATES that look just like a pile of plastic counters. It forms in dry places where lead and molybdenum ores are exposed to the air. The most brilliant wulfenite crystals were found in Red Cloud Mine at Yuma, right out in the Arizona desert. Here crystals grew so big that they could be made into gems. It's a mineral that rock hounds love to collect!

WULFENITE

WULFENITE: Group: Molybdate
Makeup: Lead molybdate
Color: Orangey-yellow • Streak: White
Luster: Glassy • Hardness: 3 • SG: 6.8

MONAZITE

KEEP WELL CLEAR OF MONAZITE—IT'S RADIOACTIVE! It's a useful mineral, though, that forms in brown-and-gold grains. It's made of phosphorus combined with one of the elements called "rare earth elements." These include cerium and lanthanum. Carmakers like rare earth elements in monazite because they are used in the catalytic converters that take some of the nasty pollutants out of car exhausts. Our smartphones rely on rare earth elements, too.

MONAZITE: Group: Phosphate
Makeup: Phosphate of rare earth metal
Color: Brown or golden • Streak: White, yellow
Luster: Resinous • Hardness: 5–5.5 • SG: 4.9–5.3

TURQUOISE

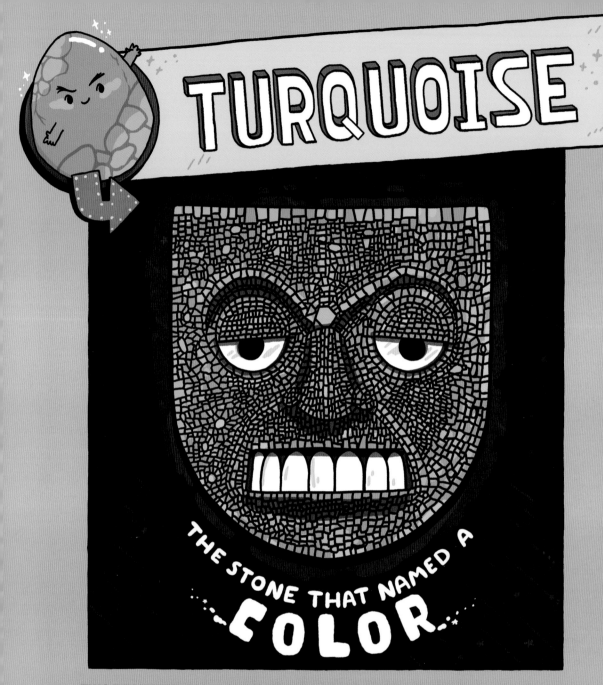

THE STONE THAT NAMED A COLOR

THIS BEAUTIFUL BLUE-GREEN STONE IS THE COLOR OF THE SEA AND THE SKY.

Native Americans call turquoise the "fallen sky stone," and believe that its color comes down from the sky in rain. It was treasured by ancient civilizations—it's in Tutankhamun's burial mask and in important Aztec masks. The most cherished shade is pale sky blue, especially if it has thin veins of impurities proving that it's entirely natural.

ANCIENT TREASURE

Turquoise gets its name from the French for "Turkey stone." That's the country, not the bird! Long ago, the best stones were mined in the deserts of Persia (now Iran), and then transported west across Turkey. Today, some of the best turquoise comes from mines in Arizona and Nevada in the southwestern United States. The ancient Egyptians probably got the turquoise for Tutankhamun's mask from the region of Mafkat, meaning "country of turquoise," in the Sinai peninsula.

You may notice what all these places have in common— they're hot and dry. That's because although it contains water, turquoise only forms in dry places. It's made of copper and aluminum and grows in lumps and veins where water seeps through aluminum-rich rocks near copper deposits. The more copper, the bluer the stone.

The Aztecs so loved turquoise that they named a god after it—Xiuhtecuhtli (pronounced "shoo-tuh-KUHT-lee"). The Turquoise Lord was the god of fire, night and day, and volcanoes.

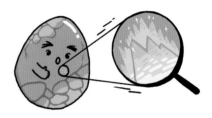

Turquoise doesn't look like it's a crystal. It looks smooth and solid. But look very close, under a microscope, and you might see some very, very tiny crystals.

You may think that buried skeletons and bones are whitish.

But chemicals turn them different colors, including turquoise color!

Hi.

TURQUOISE: Group: Hydrated phosphates • Makeup: Copper aluminum phosphate • Color: Blue-green
Streak: White with hint of green • Luster: Waxy • Hardness: 5–6 • SG: 2.6–2.8

HALITE

EDIBLE ROCK

MADE OF SALT

HALITE IS ROCK SALT, YES, THAT'S SALT IN THE FORM OF ROCK!

It's made from cubic salt crystals, and it's exactly the same salt that we add to our food. It was formed from the salt left over when seas evaporated in ancient times. Then it was buried underground for millions of years. It's usually brown but can be pink or even blue.

SALT MINES

Most rock salt was formed long ago, when places that are now land were covered by oceans. Tropical sun beat down on the water, steaming it away, leaving large salt deposits that were gradually buried. That's still happening naturally in places today like the Dead Sea, between Israel, the West Bank, and Jordan.

When we want to make food last longer and taste better, or get rid of the slippery ice on a road, we need rock salt. The ancient Egyptians even used salt to dry out a body before mummification.Underground deposits have been mined for thousands of years. The oldest-known saltworks in the world is in Xiechi Lake, in China, which probably dates back to 6000 BCE. Today, salt often comes from evaporating seawater in hot places.

Right up until the twentieth century, huge caravans of 20,000 or more camels transported salt across the Sahara desert. The ancient Romans built roads to carry salt—the most famous was the Via Salaria, in Italy.

At Wieliczka near Krakow, in Poland, there is a giant salt mine that has been mined for 700 years. There are now 150 miles (245 km) of passages where the miners have carved amazing churches and statues out of the salt.

Come and join the Roman Army. Very good _salary_!

1 MONTH LATER...

Pay day at last!

SALT

Roman soldiers were often paid in _salt_ not money, giving us the word "salary."

HALITE: Group: Salts • Makeup: Sodium Chloride • Color: White, pink, or blue • Streak: White
Luster: Glassy • Hardness: 2 • SG: 2.1+

FLUORITE

GLOWS IN THE DARK

MULTI-COLORED MARVEL

FLUORITE IS THE MOST SHOWY OF ALL MINERALS BY FAR. It comes in every color of the rainbow, and even single crystals can be a big mix of sparkling colors. It's essentially made of calcium fluoride, but its color schemes partly come from traces of rare earth elements. Yttrium, for example, gives fluorite a lovely, creamy lavender color.

FLASHY FUN

Fluorite forms gorgeous, colorful crystals. Yet it's very rarely used as a gemstone. That's because it's quite soft and brittle, and hard to cut into a shape. But collectors love it and look for prize specimens. The most prized are pinkish-purple crystals found in the European Alps, alongside smoky quartz. Tiny but wonderful little green fluorites are found in the Harz mountains in Germany, while fluorites used to be found in tin mines in Cornwall, England. And rich purple fluorites are found in the famous Elmwood mines in Tennessee.

Pure fluorite crystals are colorless. The colors come from a tiny defect in their framework of atoms called a color center, as well as impurities—traces of other elements. Some kinds of fluorite glow in the dark—this is where the word "fluorescence" comes from!

Some kinds of fluorite, such as blue john, have colors in bright, stripy bands. The name blue john comes from the French "bleu-jaune," which means "blue-yellow." That's odd because blue john is purple and cream!

Fluorite isn't just nice to look at. It can be used to make hydrofluoric acid, and that can be made into fluoride, the substance they add to toothpaste and water to keep your smile bright.

Where are the crystals?

I have a UV torch!

Fluorite glows in UV light!

FLUORITE: Group: Salts • Makeup: Calcium fluoride • Color: Many colors • Streak: White • Luster: Glassy
Hardness: 4 • SG: 3–3.3

SPINEL

TWIN TRIANGLE

✷ GREAT IMPOSTER ✷

SPINEL IS A TRICKSTER. Its red color is so similar to ruby, and its blue color is so similar to sapphire that many people have been fooled! It was once thought that a large red gem in the front of the Imperial State Crown of the United Kingdom, called the Black Prince's Ruby, was a ruby. We now know it's a spinel, one of the biggest uncut spinels in the world! No wonder they once called spinel "the Great Imposter."

POLISHED BY THE SPIRITS

Spinel forms in metamorphic rocks such as marble, in eight-sided crystals that are so perfect it's as if a jeweler cut them into shape! In Myanmar in Asia, people used to say spinels were "polished by the Spirits." Myanmar is the source of some of the best spinels, and long ago, very big spinels from here were passed by kings, queens, and emperors from one to another as spoils of war.

People used to believe that their spinels were rubies, sapphires, or even diamonds. But in 1783, French mineralogist Jean-Baptiste Louis Romé de L'Isle showed that many are a different kind of stone, which came to be called spinel. For a long time, people thought that spinel was a "fake" and didn't value it much. Now people recognize it as a lovely stone in its own right.

Spinel crystals are usually eight-sided. But sometimes they make triangles. Very rarely, two triangular spinel crystals can twin to form an amazing "Star of David."

Spinel is not a single mineral but a whole variety. There is green gahnite and black gahnite, as well as famous red "gem" spinel.

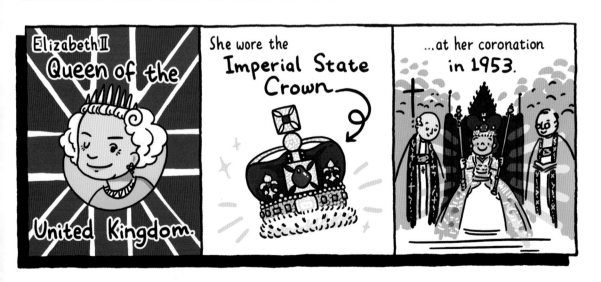

Elizabeth II Queen of the United Kingdom.

She wore the Imperial State Crown

...at her coronation in 1953.

SPINEL: Group: Complex oxide • Makeup: Magnesium aluminum oxide • Color: Typically red, but can be green or blue
Streak: White • Luster: Glassy • Hardness: 7.5–8 • SG: 3.6–4

RUBY

RARE GEM

BLOOD-RED BEAUTIFUL

CALLED RATNARAJ, KING OF GEMS, BY THE ANCIENT HINDUS, ruby is deep-red, rare, and more precious even than diamond. Rubies formed like dark cherries in a cake of marble and similar rocks, cooked in the heat of Earth long ago. But they are so tough, they survived when all the rest of the rock was worn away. And so they lie, waiting to be found by lucky people in riverbeds.

CORUNDUM COUSIN

Ruby belongs to the corundum family of minerals, which are among the world's hardest. Pure corundum is a dull brown or black. Ruby gets its amazing color from tiny impurities of chromium. It's forged deep in Earth's crust under immense heat and pressure, and is almost indestructible. Some rubies are more than three billion years old!

For centuries, the best rubies were found where they had worn out of the rock, in rivers in Myanmar in Asia. Today, one of the only ruby mines in the world is the Aappaluttoq mine, in a remote, mountainous region of Greenland. Rubies can vary in shade from almost pink to nearly purple; the most highly prized deep red color is known as "pigeon's blood."

One of the biggest rubies ever found, over four pounds (1.8 kg), was carved into the shape of the Liberty Bell. In 2011, it was stolen from a jewelry store in Delaware! There is still no trace of it!

One of the world's most expensive gems is called the Sunrise Ruby, after the name of a poem by the thirteenth-century Iranian poet Rumi. It sold at auction in 2015 for a mind-blowing $30 million!

In 1960, Theodore Maiman made the first **EVER** laser...

...out of a synthetic ruby.

RUBY: Group: Oxide • Makeup: Aluminum oxide • Color: Red • Streak: None • Luster: Adamantine • Hardness: 9 • SG: 4

MINERAL DETECTIVE

Identifying a mineral is like solving a puzzle. Sometimes you can get it from a single clue. But mostly you have to piece together a lot of clues to solve the mystery! Compare your notes to the data in this book for matches!

WHAT COLOR IS IT?

You can identify a few minerals by color alone. Beware! Impurities mean most minerals come in a variety of colors.

EASY-TO-SPOT ASSOCIATIONS

Some minerals are often found together, "assocations." It's much easier to ID them if you know which of them are friends!

- blue azurite – green malachite
- purple fluorite – black sphalerite
- red garnet – black mica
- green amazonite – smoky quartz
- brassy pyrite – milky quartz
- green apatite – orange calcite
- golden-brown barite – yellow calcite
- purple amethyst – colorless/golden calcite

YELLOWS AND GOLDS:
SHINY METAL: GOLD OR PYRITE
CUSTARD YELLOW: SULFUR
SYRUPY GOLD: ORPIMENT

REDS AND ORANGES:
BLOOD-RED: CINNABAR
CARROT: CROCOITE
DEEP RED: RUBY
RUSTY RED: JASPER
MARMALADE: CITRINE

PURPLES AND BLUES:
PALE BLACK CURRANT: AMETHYST
SEA BLUE: AZURITE, LAZURITE
SKY BLUE: SAPPHIRE

Azurite

GREENS:
FOREST-GREEN: MALACHITE
CLEAR GREEN: OLIVINE, EMERALD

HABIT: WHAT SHAPE IS IT?

You can't always see perfectly shaped crytals, but you can often spot a particular mineral by the distinctive shape it forms as it grows, known as its "habit." You can identify the minerals by looking at pp. 8-9.

...dlelike ...sters

Slender, bladelike crystals

Rounded, like grapes

Branches like a tree

Sugary Coat

...y thin crystals ...e fibers

Layered like leaves

Ball-shaped

Masses, with no Clear shape

Kidney-shaped

Needles inside

Flat crystals, like tabletops

LUSTER: WHAT DOES IT LOOK LIKE?

Hard and super shiny

Metallic

Pearly

Like glue

Silky

Serpentinite

Waxy

Vitreous

MINERAL DETECTIVE

You can do three simple tests at home to confirm a mineral's identity.

STREAK! Try scraping your mineral across the unglazed back of an old porcelain tile. Many minerals will leave a colored streak that's always the same color.

Jade

Calcite

Barite

Opal

Bauxite

Gypsum

Emerald

Siderite

Malachite

Wulfenite

Beryl

Spinel

Galena

Orpiment

Monazite

Tourmaline

Bournonite

Realgar

Chalcedony

Spodumene

Pyrite

Turquoise

Quartz

Magneti[te]

Limonite

Talc

Halite

Lazurite

Cinnabar

Gold

Crocoite

Fluorite

A white streak can be one of many minerals, but you know it's not one with a colored streak, or no streak at all.

Hematite

Sulfur

Aragonite

Feldspar

TOUGH TESTING

The hardness of minerals is measured on the Mohs scale, in comparison to ten reference minerals. You can work out where your mineral is on the scale by finding out which of the reference minerals it will scratch and which are scratched by it. You'll need a sample of each mineral in the scale, or one of the stand-ins suggested.

MOHS SCALE

 1. TALC

 2. GYPSUM or a finger-nail

 3. CALCITE or a nickel

 4. FLUORITE or an iron nail

 5. APATITE or an old drinking glass

 6. FELDSPAR

 7. QUARTZ

 8. TOPAZ or emery sandpaper

 9. CORUNDUM

 10. DIAMOND

WEIGHING IN

You can check out a mineral's specific gravity (SG)—its density compared with water. You'll need a spring balance, which you can get from a hardware store.

1. Hang your sample on the balance and take a measurement.

2. With the mineral still hanging, immerse it in a bowl of water. Take another measurement.

3. The SG is your first measurement, divided by the difference between the two measurements.

HEMATITE

Hematite is 70 percent iron, and the world's main source of its most important metal. It often comes in knobby lumps shaped like kidneys, called kidney ores, but it's also spread in layers through sedimentary rock beds. And it's in red ocher, used by artists to make the very first cave paintings, tens of thousands of years ago.

HAEMATITE: Group: Oxide
Makeup: Iron oxide
Color: Steel gray to earthy red
Streak: Red
Luster: Metallic
Hardness: 5–6
SG: 5.3

MAGNETITE

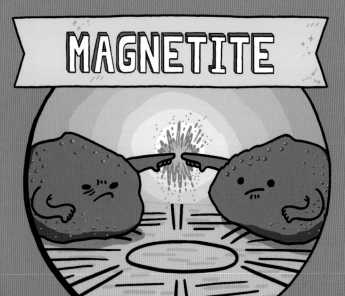

Magnetite is magnetic, the most naturally magnetic mineral on Earth. Legend has it that a shepherd in ancient Greece discovered magnetism when the nails in his shoes stuck to stones packed with magnetite. Later on, sailors used magnetite stones, called lodestones, to make the first compasses, to help them find their way across the sea. Like hematite, magnetite is also a main ore of iron. You have a tiny amount of magnetite in your brain!

MAGNETITE: Group: Oxide
Makeup: Iron oxide
Color: Black
Streak: Black
Luster: Metallic
Hardness: 5.5–6.5
SG: 5.1

SIDERITE

Siderite is a much softer iron ore than hematite or magnetite. It's found in thin beds in sedimentary rocks such as shale, clay, or coal. Sometimes, it mixes with clay to form beds almost as hard as iron, called clay ironstone. Nodules of clay ironstone are often packed with tiny fossils of millipedes and clams.

SIDERITE: Group: Carbonate
Makeup: Iron carbonate
Color: Dark Brown
Streak: White
Luster: Glassy
Hardness: 3.5–4.5
SG: 3.9+

LIMONITE

When iron minerals rust on a surface, and stay damp, you can get yellow-brown limonite. The yellow in many cave paintings is powdered limonite—yellow ocher. Limonite is found in huge, ancient beds up to 2,000 feet (60 m) thick called banded iron formations (BIFs). These formed early in Earth's history when the oceans were rich in iron. There's a famous BIF near Lake Superior.

LIMONITE: Group: Oxide
Makeup: Hydrated iron oxide
Color: Yellow, brown
Streak: Brownish-yellow
Luster: Earthy, dull
Hardness: 4–5.5
SG: 2.9–4.3

BAUXITE

BAUXITE IS THE MAIN SOURCE OF THE WORLD'S ALUMINUM. It's a very soft browny-white earth mix of minerals. It's weathered into thick layers in warm, damp, tropical conditions. Aluminum is a fantastically useful metal: light, tough, resistant to corrosion, and used for everything from soft drinks cans to airplanes. It's also infinitely recyclable: 75 percent of aluminum ever used is still in use today!

COSTING EARTH

When French geologist Pierre Berthier discovered a red earth near Baux in France in 1821, no one was that bothered. They had no idea that bauxite is absolutely jam-packed with aluminum in the form of aluminum hydroxide, as well as other minerals.

In 1824, Danish scientist Hans Oersted managed to extract some pure aluminum metal for the first time. For a while, it was still very rare and special. Then, in 1886, it became possible to get lots of aluminum from bauxite by zapping it with electricity, and that bauxite could be found in huge layers around the world. Now aluminum is the world's most used metal after iron, and almost all of it comes from bauxite.

The world uses 180 billion aluminum drinks cans a year! Aluminum is 100 percent recyclable, so remember to recycle empty cans!

In the 1850s, aluminum was still so rare that French Emperor Napoleon III's top dinner guests dined off aluminum plates, while ordinary guests slummed it on gold and silver. And his baby's rattle was made from aluminum and gold!

Bauxite to Aluminum

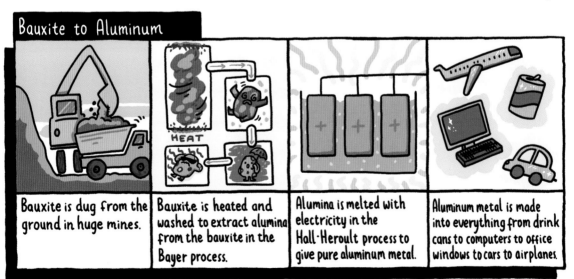

Bauxite is dug from the ground in huge mines.

Bauxite is heated and washed to extract alumina from the bauxite in the Bayer process.

Alumina is melted with electricity in the Hall-Heroult process to give pure aluminum metal.

Aluminum metal is made into everything from drink cans to computers to office windows to cars to airplanes.

BAUXITE: Group: Oxide • Recipe: Aluminum hydroxide • Color: Reddish, buff, silvery-white • Streak: White • Luster: Metallic
Hardness: 1.5 • SG: 2.72

CALCITE

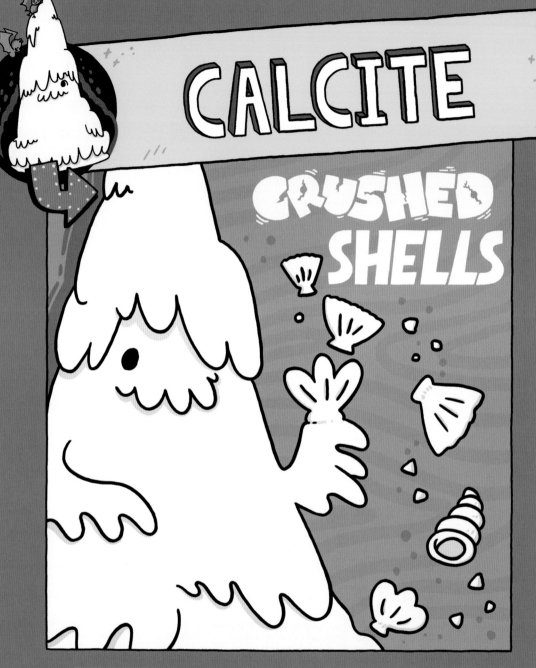

CRUSHED SHELLS

EVERYWHERE YOU SEE LIMESTONE ROCK—IT'S MOSTLY CALCITE.
When you see furring in kettles and around the bathtub—that's calcite, too!
It forms 300 different kinds of crystals, with names such as dogtooth spar
and nailhead spar because of their distinctive shapes.

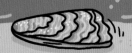

SHELLS AND BONES

Calcite columns stack side by side to make an eggshell.

A whole lot of the world's calcite is shellfish! That's because shellfish shells are made of calcite and a similar mineral called aragonite. Limestone rock (see pp. 32–33) is mostly calcite, made mostly of the shells and skeletons of tiny sea creatures that piled up over millions of years.

Calcite is chemically calcium carbonate, a combination of calcium metal, carbon, and oxygen. Shellfish get calcium from the water around them. Amazingly, there are over 550 trillion tons (500 trillion tonnes) of the metal calcium dissolved in the oceans! There's a lot of carbon dioxide that's dissolved, too. Shellfish take these components from the water and bind them together to make their shells. We get the calcium we need from dairy foods, such as milk and cheese, and from leafy green vegetables, such as kale and broccoli.

Iceland spar is clear crystals of pure calcite that shows a double image when you look through it. It was probably found by the Vikings and has special properties that helped them navigate the oceans.

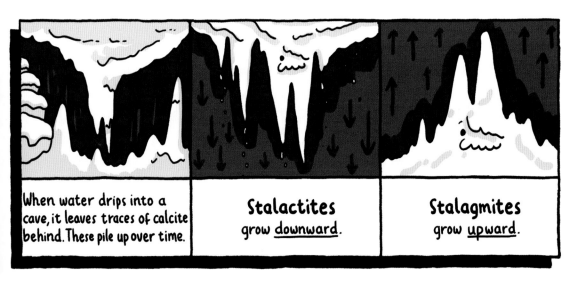

| When water drips into a cave, it leaves traces of calcite behind. These pile up over time. | Stalactites grow <u>downward</u>. | Stalagmites grow <u>upward</u>. |

CALCITE: Group: Carbonate • Makeup: Calcium carbonate • Color: White • Streak: White • Luster: Glassy
Hardness: 3 • SG: 2.7

ARAGONITE

SHELLS AND PEARLS...

ARAGONITE IS CALCITE'S TWIN. It's chemically identical to calcite, but its crystals are different shapes, and it's tougher and denser. Alongside calcite, it's in the shells of shellfish, and it's the mineral that millions of coral polyps build into skeletons to make a coral reef.

JEWELS OF THE SEA

Inside many rough-looking shellfish shells, you'll find a super-smooth lining that shines and shimmers silvery-white with rainbow colors. This magical lining is called "nacre" or "mother of pearl" and it's the shellfish's protective inner coat.

Oysters can make nacre into beautiful little balls called pearls, which are used as gems. Pearls can be black, rose red, blue, green, purple, yellow, or white. The most valuable are white and silvery-white saltwater pearls.

Nacre is mostly aragonite. But it's bound together with a hornlike material called conchiolin. The nacre's shimmering rainbow colors come from the way layers of conchiolion and aragonite interfere with light and split it into different colors.

Pearl divers dive down to the sea floor to look for oysters that might hold a beautiful pearl.

Iron flowers, known as flos ferri, seem to bloom out of iron ore. But they're not flowers at all—they are white crystals of aragonite that grow slowly into crazy shapes.

When a grain of sand gets into an oyster...

It oozes nacre to wrap around the sand...

Layers of nacre build up.

Now there's a little ball: a pearl!

ARGONITE: Group: Carbonate • Makeup: Calcium carbonate • Color: White • Streak: White • Luster: Glassy
Hardness: 3.5–4 • SG: 2.9–3

MALACHITE

COPPER TOP

GREEN QUEEN

MALACHITE IS STRIPY, SHIMMERING GREENS THAT NEVER FADE.
It's the velvety-green residue that forms when copper, bronze, or copper-rich rocks are exposed to air. It forms in big, glossy, rounded lumps in caves and cavities when water rich in copper minerals trickles through the ground. When sliced, the lumps show wonderful, curvy layers.

GOING GREEN

Long ago, people realized that malachite was a glowing green signpost in the ground to show the way to deposits of copper—the first metal used by humankind. But the ancient Egyptians realized it also looks stunning in and of itself when carved and polished. Malachite was one of the first green pigments to be used in painting because it can easily be ground into a fine powder. The ancient Egyptians used it in their tombs.

In the 1700s, huge deposits of malachite were found under the Ural mountains in Russia. The Russian royal family loved it, and in St. Petersburg, the Winter Palace has a gorgeous green room where many things are made of glowing green malachite. Now most malachite comes from the Democratic Republic of the Congo.

When there's a little more water around, malachite sometimes turns into a vivid blue stone called azurite. Hundreds of years ago, some painters preferred azurite blue to malachite green.

In the Ural mountains of Russia, legend has it that the copper deposits are guarded by a spirit, the Malachite Queen.

Leonardo's colors were made from:

White
Lead that is left to flake in horse manure.

Red
Crushed cinnabar found at a volcano.

Green
Malachite ground to a powder.

Vivid Blue
Azurite ground to a powder.

Soft Blue
Crushed lapis lazuli.

MALACHITE: Group: Carbonate • Makeup: Copper carbonate hydroxide • Color: Green • Streak: Pale green
Luster: Glassy to dull • Hardness: 3.5–4 • SG: 4

SANDY PLANET

OUR WORLD IS BUILT ON SAND! Sandy minerals, called silicates, are the most abundant of all minerals. Nearly half of all minerals are silicates, and igneous rocks, which make up 90 percent of Earth's crust, are mostly made of silicates.

FELDSPARS

LAZURITE

QUARTZ

CITRINE

AMETHYST

Making silicates: All silicates are made from the elements silicon and oxygen, along with a metal element or two, or maybe even three! They are tough and long-lasting, and survive long after the rock in which they formed gets recycled into new rocks, again and again. When you see sand on the beach, it's mostly silicates in between its time as rocks! Some silicates get made or remade in sedimentary and metamorphic rocks. Most, though, started off when magma cooled to make igneous rocks.

Diatoms: The hard bits of bones are mainly made with calcium. But microscopic sea organisms, called diatoms, have skeletons that are silica. When they die, they pile up on the seafloor as a sandy slime, which sometimes forms new rocks.

Silicate pyramids: Scientists divide silicates into families according to the shape their atoms make when they join to make molecules. The simplest is a pyramid with a silicon atom with four oxygen atoms. All the other shapes are built from this.

TIGER'S EYE

CHALCEDONY

JADE

TOURMALINE

BERYL

FELDSPARS

ROCK FORMERS

←ALUMINUM AND SAND

FELDSPARS ARE EVERYWHERE! Two-thirds of Earth's crust is feldspars, and almost every other mineral is just filling in the gaps. They're made of aluminum silicate (see pp. 88–89) minerals. Feldspars don't often make pretty gems, but they're the unsung heroes in our homes, helping to make everything from mugs to bathroom tiles.

MOST COMMON

Feldspars are just about everywhere because they crystallize out of magma, the molten material that comes up from the deep to form the rocks of Earth's crust. You can also find them in rocks in space, including meteorites!

There's a whole bunch of feldspars, split into two kinds. Firstly, there are feldspars rich in potassium, such as orthoclase and sanidine. Igneous rocks like granite are mostly made of potassium-rich feldspars. So are metamorphic rocks like gneiss. Then there are "plagioclase" feldspars, such as albite and anorthite, which are rich in sodium or calcium. These feldspars make rocks such as gabbro and make the moon white.

Moonstone is a feldspar with a milky sheen. In India, it's said if you hold it carefully in your mouth during a full moon, you can see your future. But don't try this until you're an adult.

Labradorite shimmers like a stained-glass window. A legend of the Inuit people of Canada says that it fell from the icy fire of the aurora borealis, the lights that in the night sky appear over the Arctic.

About 2,000 years ago, Chinese people hardened clay with a secret ingredient, the feldspar petuntse.

Chinese porcelain is **beautiful.**

FELDSPARS: Group: Silicates • Makeup: Aluminum silicate • Color: Off-white • Streak: White • Luster: Glassy
Hardness: 6–6.5 • SG: 2.53–2.76

LAZURITE

BLUE AS NIGHT

MOUNTAIN MAGIC

LAZURITE IS A SOFT SILICATE MINERAL OF DEEP, VIVID BLUE—WHEN IT'S EMBEDDED WITH DOTS OF GOLD PYRITE, IT LOOKS LIKE A STARRY MOONLIT NIGHT. It's the major ingredient in the blue gemstone lapis lazuli and forms in veins and masses in marble. The deep blue comes from sulfur. By itself, sulfur is bright yellow, but when it joins with the other chemicals in lazurite, it turns the mineral a blue color.

BEAUTIFUL BLUE

The vivid blue color of lapis lazuli totally entranced people in the ancient world. It appears in the Bible and in the Sumerian epic of Gilgamesh, one of the world's oldest poems, dating from almost 4,000 years ago. The ancient Egyptians loved it. It's in the mask of the boy pharaoh Tutankhamun, making a wonderful contrast with the gold.

You can tell how much everyone liked it, because you could get it from only one place back then, so you had to make a major effort to get it! That place was faraway Sar-e-Sang in the Kokcha Valley, high up in the mountains of Afghanistan. Today lapis lazuli is also mined near Lake Baikal in Siberia, Russia, and in Ovalle in Chile. But it's still very rare.

Not all the blue in Tutankhamun's mask is lapis lazuli. It was too expensive even for royalty to use a lot. So the blue stripes are actually glass paste dyed blue. Only the eyes and eye surrounds are real lapis!

A birthstone is the gem connected to the month you're born. Lapis lazuli is a traditional birthstone for December.

Painters like Leonardo da Vinci, in the 16th century, loved the deep blue paint Ultramarine.

It was made from crushed **lapis lazuli.**

It was more precious than gold!

LAZURITE: Group: Silicates • Makeup: Sodium calcium aluminum silicate sulfur sulfate • Color: Off-white • Streak: White
Luster: Glassy to greasy • Hardness: 5.5–6 • SG: 2.6

QUARTZ

After feldspars, quartz is the most abundant mineral in Earth's crust. When magma is quite silica-rich, as it cools, there will be some left at about 750–800°C (1,382–1,472°F). From this quartz forms (as well as feldspars and maybe muscovite mica). When other minerals in rocks are turned to dust by weathering, quartz toughs it out as grains and pebbles. The pebbles, gravel, and sand on the beach are mostly quartz. And sedimentary rocks, such as sandstone, are rich in quartz—because quartz grains survive in sediments to make up the new rock. Citrine, amethyst, and tiger's eye are varieties of quartz.

QUARTZ: Group: Silicates • Makeup: Silicon dioxide
Color: Clear, or just about any color
Streak: White • Luster: Glassy
Hardness: 7 • SG: 2.65

CITRINE

Quartz is a shape-shifter! A tiny change—the trace of an element, heat or radiation, the bubbles of a gas—turns it into something different. Citrine is quartz made yellow by traces of iron oxide. It's the most precious of all the quartz gems. In ancient times, people carried citrine to protect them against snake bites and evil thoughts! Some amethyst turns to citrine when it comes very close to hot magma. Today, most citrine gems you see for sale were created by heating amethyst.

AMETHYST

Amethyst is often found lining geodes. The largest amethysts come from giant geodes in Brazil and Uruguay. Scientists will tell you that amethyst is quartz colored purple by traces of iron. But ancient Greek legend tells a different story. The goddess Artemis turned a girl, Amethystos, into a crystal statue to save her from wild tigers let loose by Dionysus, god of wine. Grief-stricken Dionysis poured wine filled with his tears onto the statue, turning it purple.

TIGER'S EYE

Tiger's eye is a quartz variety that's brown or amber with yellow stripes, created by traces of iron oxide. Sometimes it has a bright band across the middle that almost glows. That's what makes it look like a tiger's eye glinting! This effect, called a cat's-eye effect, is caused by the way parts of the crystal grow like fibers. A similar effect is seen in the rare blue-green quartz crystal, hawk's eye.

CHALCEDONY

MICROSCOPIC CRYSTAL

PERFECTLY POLISHABLE

SOME KINDS OF QUARTZ LOOK LIKE BEAUTIFUL SMOOTH STONES, not like angular crystals at all. In fact, they're made from great masses of crystals, but the crystals are so tiny they're virtually invisible. They can be cut and polished into beautiful stones. They're called chalcedony and include incredible gems such as agate, carnelian, jasper, chrysoprase, and onyx.

CHALCEDONY: Group: Silicates • Makeup: Silicon dioxide • Color: Clear, or just about any color
Streak: White • Luster: Glassy • Hardness: 7 • SG: 2.65

CHALCEDONY GEMS

Agate is a super-stripy mineral that comes in many colors, and looks a little like candy! It forms in layers, hardening inside liquid bubbles in rock. The layers show as stripes when the bubble is cut open. Agate is very tough and can make a cutting edge, but also makes lovely jewelry. Agate comes in many varieties, including thundereggs, hollow stones with frizzy crystals inside that look a little egglike when cracked.

When you look at a beautiful agate, you are looking back in Earth's history. It has taken as much as 50 million years to form.

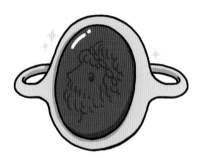

The red-brown gem carnelian gets its color from traces of hematite (see p. 78). It's translucent, which means it lets a little light in, so it seems to glow. Ancient warriors wore carnelian around their necks for courage and for the power to vanquish their enemies. In ancient Egypt, it was known as the setting sun, and master architects wore carnelian to show their importance.

From ancient times until the nineteenth century, people carved their special sign onto a carnelian ring, then pressed the ring into hot wax to seal a letter.

JADE

O HARD AND GREEN

MICROSCOPIC CRYSTALS

JADE IS A VIVID GREEN, GLASSY MINERAL FORMED IN SUPER-HOT LAVA.
It has been treasured for thousands of years for its toughness and beauty—hard enough to make knives, gorgeous enough to make jewels. It's best known for being carved into little statuettes in China, such as dragons, lions, and dogs. These carvings require incredible skill and last forever.

DOUBLE JADE

Jade comes in two kinds. One is the mineral jadeite; the other is the mineral nephrite. Jadeite is harder, denser, and more translucent. It's also rarer and more valuable. Nephrite is softer, less dense, and more opaque.

All the jade used in the oldest Chinese statues is nephrite. That's because jadeite can't be found in China. The Chinese only started to use jadeite when they imported it from Myanmar after the eighteenth century and discovered imperial jade, an almost clear jadeite, which gets its stunning emerald color from traces of chromium.

Even in Myanmar, jadeite's not so easy to find. In fact, except in Guatemala, you will never find jadeite in the original rock where it formed. Instead, if you're lucky, you'll find it as water—worn pebbles in stream beds, where it was washed after being weathered out of rock.

About 2,000 years ago, wealthy Chinese people got entire suits of armor made from squares of jade, held together by gold, silver, or copper wires if you were royal and silk threads if you weren't. The armor wasn't for fighting in. Instead, when people died, they'd be buried in their jade armor to keep away bad spirits.

The Olmec, Mayan, and Aztec peoples of ancient South America loved jade.

Some thought it could cure pain if you held it to your side.

Jade means "stone of the side."

JADE: Group: Silicates • Makeup: Sodium aluminum iron silicate (jadeite) • Color: Green, white, or yellow • Streak: White
Luster: Glassy • Hardness: 6.5–7 • SG: 3.25–3.35

TOURMALINE

ANY COLOR YOU LIKE

MAD MIX OF CHEMICALS

TOURMALINE IS THE MOST WONDERFULLY COLORFUL MINERAL OF ALL. Actually, it's made from a whole stew of minerals and comes in dozens of varieties. But you name a color, tourmaline turns up in it. The ancient Egyptians believed it soaked up all the colors of the rainbow, and called it rainbow rock.

ELECTRIC BLUE

Traces of different elements turn tourmaline into more than 100 different color varieties, including the best gem variety, elbaite, named after the Italian island of Elba. But in recent years, the variety everyone wants is a kind of elbaite called Paraiba. Paraiba is electric blue, or violet—more like a neon light than a gem! It gets its startling blue color from traces of copper, while a bit of manganese in addition turns it purple.

Paraiba was first found in 1989 in pegmatite pockets in Paraiba and Rio Grande do Norte in Brazil. People just loved it, and soon chunks of Paraiba were selling for up to $50,000. Wow! In 2001 and 2005, similar gems were found in Nigeria and Mozambique. They look the same, but chemical analysis shows they're different. So is this African stone Paraiba? Some say yes; others say no.

If conditions change as they grow, tourmalines can grow with different color bands. Crystals like these are called "zoned crystals." Pink-and-green "watermelon" tourmaline looks like a slice of watermelon!

TOURMALINE: Group: Silicates • Makeup: Sodium lithium aluminum boro-silicate hydroxide • Color: Pretty much any color, especially black or blue • Streak: White • Luster: Glassy • Hardness: 7.5 • SG: 3–3.2

BERYL

CRYSTAL GIANT

BERYL IS A GEMSTONE THAT COMES IN LOTS OF DIFFERENT VARIETIES.
It grows in pegmatites—igneous rocks that form out of magma that has almost
completely cooled—treasure houses of large minerals. It also forms in vugs—cavities
found in granite, in a rainbow range of colors! Most beryl crystals are small,
but in certain conditions, they can grow as big as telephone poles!

COAT OF MANY COLORS

Beryl just loves changing color! It only needs a little trace of a metal to make the switch. Chromium and vanadium turns beryl to green emerald. Traces of iron changes clear beryl into beautiful sea-blue aquamarine, and a different kind of iron adds yellow to aquamarine to make it greenish-blue. A tiny amount of iron creates heliodor, a lovely yellow beryl whose name means "gift of the sun." Meanwhile, the metal manganese (or sometimes cesium) makes pinkish morganite and the super-rare "red" beryl, which can look like ruby.

What's interesting is that these colors are not fixed. Heating the gems can intensify the color. In fact, many aquamarines you see in jewery shops didn't come out of the ground quite as bright blue—they were cooked!

Beryl is an ore of beryllium—a light, tough metal used for spaceships.

Morganite was discovered in Madagascar in 1910 by the chief gemologist for the famous jewelry store Tiffany & Co. It was named after Tiffany's best customer, the banker J. P. Morgan.

In the old tin mines of Cornwall, England, the miners found little holes in the rock where amazing crystals like beryl grew. They called them "vugs" after the Cornish vooga, meaning cave.

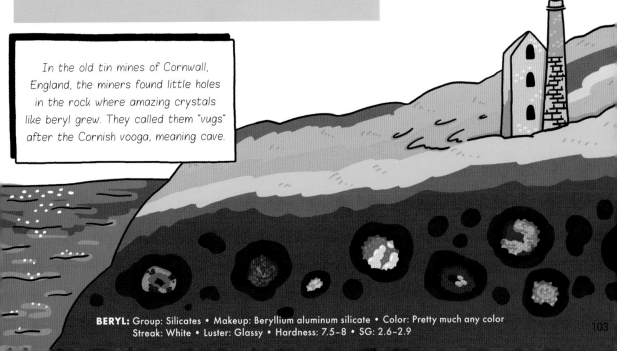

BERYL: Group: Silicates • Makeup: Beryllium aluminum silicate • Color: Pretty much any color
Streak: White • Luster: Glassy • Hardness: 7.5–8 • SG: 2.6–2.9

EMERALD

THE GREEN GODDESS

ONLY FOUR GEMS ARE CALLED "PRECIOUS" FOR THEIR QUALITY, RARITY, AND BEAUTY—DIAMOND, SAPPHIRE, RUBY, AND EMERALD. (All other gems are called "semiprecious.") For many people, emerald is the most precious gem of all. The other precious gems come in various colors, but emerald is very particular, and comes only in vivid, glistening green. Emerald is a very special form of beryl and gets its green color from small traces of chromium and sometimes vanadium.

OLD GREEN

Emeralds are survivors. The oldest are nearly three billion years old. Some were made long ago where steamy waters oozed up into metamorphic rock. Others formed in bubbles of hot fluid left as magma cooled into pegmatites. Because emeralds form deep inside rock, they have to be mined. But if you're exceptionally lucky, you might find one in a stream where it was washed after being worn out of the rock over millions of years.

Cleopatra, a queen of ancient Egypt, loved emeralds and had them mined near the Red Sea. They were also much loved in the empire of the Incas in South America. Now Colombia is the center of emerald mining.

The world's biggest emerald is the Bahia emerald, weighing 752 pounds (340 kg)! It was dug up in Brazil in 2001, then almost lost in Hurricane Katrina, in 2005, while stored in New Orleans.

Most emeralds contain tiny imperfections called jardin from the French for "garden," maybe because they look like tiny plants. The emerald's jardin make it easier to break!

The Lost Emerald

In 1622, a Spanish galleon sailed from Cuba with 70lbs (32kg) of emeralds.

Two days later, it sunk.

In 1985, divers recovered an emerald said to be worth $5million.

It got stolen in 2016 and has never been found.

EMERALD: Group: Silicates • Makeup: Beryllium aluminum silicate with chromium or vanadium • Color: Green
Streak: White • Luster: Glassy • Hardness: 7.5–8 • SG: 2.6–2.9

SMOOTH AS SILK

RAINBOW DROPS

OPAL IS UNLIKE ANY OTHER GEM. IT'S NOT CRYSTAL; IT'S LIKE HARD GELATIN.

It's typically pearly white or black, but a kind called "precious opal" has a rainbow shimmer of colors called opalescence. Opal's color is never fixed, but changes continually as the light changes, or you look from a different angle.

WIGGLE WIGGLE

FAIRY LIGHTS

Only the rare, precious, dewdrop-shaped opals are used as gems. They form as tiny globules in lava, which are shaped into cushion-shaped "cabochons" by jewelers. The globule is made up of lots of tiny spheres, and it's the way light bounces through the spheres that gives opal its beautiful rainbow colors.

You can sometimes see slivers of opal in fossils or on ancient bits of wood. Crusts and lumps of opal often form in cracks and veins in rocks, too, but this is mostly plain common, or potch, opal, which is mined for use as an abrasive or filler.

Coober Pedy in the scorching Australian outback is the opal capital of the world, where the gems are mined from sedimentary rocks like sandstone and ironstone.

In the Middle Ages, some people thought that if you wrapped an opal in a bay leaf it would make you invisible.

Ancient Australian stories tell how the Creator came to Earth on a rainbow, and where his foot touched, the stones came alive with all the colors—the birth of opals.

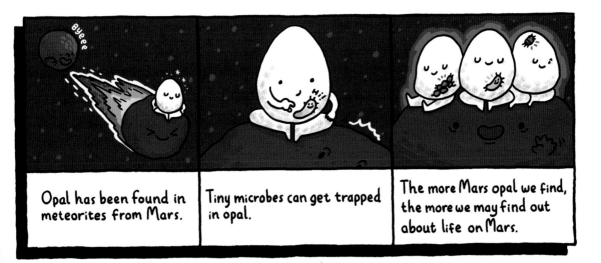

Opal has been found in meteorites from Mars.

Tiny microbes can get trapped in opal.

The more Mars opal we find, the more we may find out about life on Mars.

OPAL: Group: Silicates • Makeup: Silicon dioxide with water • Color: White, rainbow • Streak: White
Luster: Glassy • Hardness: 5.5–6 • SG: 1.8–2.3

TOPAZ

SUPER TOUGH

CLEAR AND YELLOW

TOPAZ HAS CONFUSED PEOPLE FOR CENTURIES. Until about 200 years ago, it was thought that all yellow gems were topaz. But although topaz is often yellow, it can be clear, blue, or many other colors. It's one of the hardest of all gems, and clear topaz is often mistaken for diamond.

FIRESTONE

Topaz has a fiery history. It may have got its exotic name from the ancient Sanskrit word, tapaz, for fire. Or from an island in the Red Sea caled Topazios, where it may have been mined in ancient Greek and Roman times. Ancient Egyptians believed that topaz was colored with the golden glow of the sun god, Re.

Today, topaz is found in mountainous areas around the world, especially in Brazil. It grows from mineral-rich fluids in cracks and veins in granite, and in pegmatites (see p. 17). The fluid has to contain at least a tiny amount of fluorite. One of the rarest gems in the world, the American golden topaz, has 172 facets and weighs 22,892.5 carats (10.09 lb or 4.58 kg). Beautiful blue topaz is the state gemstone of Texas.

Imperial topaz is the most rare and beautiful topaz of all. The emperors of Russia, the czars, took any that they knew about!

In the United States, you can go hunting for topazes on Topaz Mountain, in the Thomas Range of Utah.

In 1984, a gigantic yellow topaz was found in Brazil!

It's the world biggest cut gem— 13.6lbs (6.7kg).

It's as heavy as a big baby!

TOPAZ: Group: Silicates • Makeup: Aluminum flourohydroxysilicate • Color: Colorless, yellow, blue, pink • Streak: White Luster: Glassy • Hardness: 8 • SG: 3.5–3.6

GARNET

Cherry Stone!

GARNETS ARE VERY TOUGH AND VERY BEAUTIFUL. They were forged long ago, deep in Earth, often in peridotites (see p. 17) that were dragged to Earth's surface in volcanic eruptions. They look a little like dark cherries in a Black Forest cake.

NOAH'S GUIDE

Though red is the best-known color for garnet, it's actually a whole jumble of minerals that are mostly green and gray. So a garnet can actually look more like a pea than a cherry.

All these garnets form under immense pressure and heat, and there are more than 20 varieties, with some pretty weird names, such as grossular, uvarovite, and—very confusingly—topazolite. And then, of course, there is . . . demantoid! It sounds like something out of a Harry Potter book, but it's actually the most sought after garnet of all. Demantoid's green, like emerald, and sparkles as brilliantly as diamond, which is how it got its name. In the Bible story about Noah's ark, Noah is said to have a garnet lantern to light the way through the flood.

In the past, tiny garnets were used as watch bearings because of their toughness and durability.

Grossular garnet isn't gross at all. The name grossular means "like gooseberries," and it gets its name because the green variety looks like gooseberry jam. Sweet!.

Garnet Species:

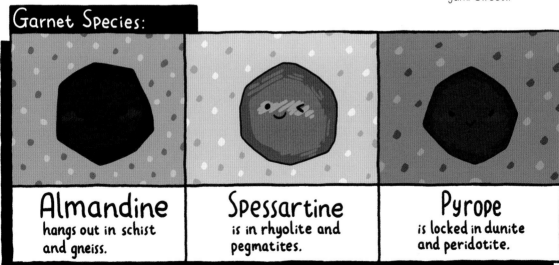

Almandine
hangs out in schist and gneiss.

Spessartine
is in rhyolite and pegmatites.

Pyrope
is locked in dunite and peridotite.

GARNET: Group: Silicates • Makeup: Calcium iron silicate • Color: Green, gray, or red • Streak: White
Luster: Glassy • Hardness: 6.5–7.5 • SG: 3.8

SPHENE AND SPODUMENE

SPHENE AND SPODUMENE ARE SUPER-DESIRABLE MINERALS.

But they only form in very rare conditions, where titanium or lithium teams up with sandy silicate. Both are found in pegmatites (see p. 17), but the best yellow sphene crystals occur in cracks in schist in the Alps mountain range in Switzerland and Austria, in Europe.

SPHENE

Sphene gets its name from the Greek word for wedge, because of its distinctive wedge-shaped crystals. It's also known as titanite because of its titanium content. This stone is seriously flashy! It has the most "fire" (internal sparkle) of any gem—even more than diamond. And it typically flashes not clear but in greens and yellows.

Sphene is mined in Canada and Mexico, and on the island of Madagascar off the coast of southern Africa.

Going out in the sun tomorrow? Sphene is the main source of titanium dioxide, the white powdery pigment they put in sunscreen.

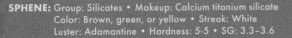

SPHENE: Group: Silicates • Makeup: Calcium titanium silicate
Color: Brown, green, or yellow • Streak: White
Luster: Adamantine • Hardness: 5-5 • SG: 3.3–3.6

Some spodumene crystals are as big as logs. The biggest ever spodumene crystals come from the Black Hills of South Dakota, where some grow up to 50 feet (15.2 m) in length!

SPODUMENE

Spodumene gets its name from the Greek for "burned to ashes" because of its gray ashen color after it's ground up for industrial use. But it makes lovely, tiny, pink gems called kunzite, too. It's also one of the main sources of lithium, the rare element that powers the batteries in countless electronic devices, such as phones and electric car batteries. We now need lithium for our batteries so badly that people are scouring the world for spodumene.

SPODUMENE: Group: Silicates • Makeup: Lithium aluminum silicate
Color: Grayish-white plus pink or yellow
Streak: White • Luster: Glassy
Hardness: 6.5–7 • SG: 3–3.2

TALC AND MICA

BIG SOFTIE

BABY POWDER

BIG FLAKE

TALC AND MICA ALMOST NEVER FORM NICE, COLLECTIBLE CRYSTALS.
They're just in thin layers that crumble apart. But because they're both useful minerals, it's worth knowing about them. Talc is super soft, the softest mineral of all. It forms when limestone is cooked by magma. Mica is super flaky and is part of the magma mix making igneous rocks.

TALC

Though it's soft, talc is one of the most universally useful of all minerals. It's used in everything from papermaking—to bind the fibers—to ceramics, paints, roofing materials, plastics, rubber, the catalytic converters of cars, wire and cable insulation, chewing gum, hoses, vinyl flooring . . . Phew!

Talc is the key ingredient in a solid rock called soapstone that forms when rocks such as peridotite and dunite get metamorphosed (changed by heat and pressure). Soapstone's so soft that it's long been used for carving.

Talc is the standard for grade 1 in the Mohs scale of hardness (see p. 77). It's so soft you can scratch it with your fingernail.

TALC: Group: Silicates • Makeup: Magnesium silicate hydroxide
Color: White • Streak: White • Luster: Dull
Hardness: 1 • SG: 2.7

MICA

oh hi...

Mica may be flaky, but it's not soft, like cornflakes. There are more than thirty different kinds that are mostly dark brown and filmy, but very tough and resistant to heat.

Two of the best-known kinds of micas are biotite and muscovite. Biotite is the glittery black "pepper" bits in granite. Muscovite mica is used to provide insulation for electrical installations. Ground up, it is used to make artificial snow for Christmas trees—yes, it grinds up white, just like its streak!

Mica ends up everywhere. If you've got walls, you've probably got mica in them. Mica is ground up and added to the plaster they make plasterboard walls and ceilings with.

MICA: Group: Silicates • Makeup: Aluminum silicate hydroxide
Color: Black, brown, white • Streak: White
Luster: Glassy to pearly • Hardness: 2.5 • SG: 2.9

GOLD AND COPPER

THE FIRST AND THE MOST IMPORTANT

LIKE GOLD, COPPER HAS ITS OWN, UNMISTAKABLE COLOR AND IS FOUND IN THE GROUND AS A PURE METAL. That's why copper and gold were probably the first metals humans ever used, since it took them a while to realize other metals could be melted out of rock. While gold is mineral royalty, it's actually copper that powers our world.

GOLD

Gold is one of the few native minerals to be found naturally in rocks in a pure state. It never corrodes, staying yellow and sunshiny forever. That's why people have always made crowns and rings and coins from it. But it's also a great conductor, which is why it's used in electronic devices like phones.

Gold is sometimes found in grains and nuggets in river gravel and can be sifted out. But all this river gold comes originally from veins in rocks, where gold is often found with white quartz and sulfide minerals such as stibnite. About 190,000 metric tons have been mined, and it nearly all still exists.

In 1869, two miners in Australia found the biggest gold nugget ever, bigger than a pumpkin. It yielded 56.6 pounds (71 kg) of gold! The nugget was named "Welcome Stranger."

GOLD: Group: Native element • Makeup: Gold • Color: Gold • Streak: Gold • Luster: Metallic • Hardness: 2.5–3 • SG: 19.3

COPPER

Copper is a superconductor. Like silver, it conducts heat and electricity fantastically well. That's why most of the electrical wires that power our homes are made from copper. Germs hate copper! Its electric properties make it great at killing bacteria, viruses, and fungi. That's why more and more hospitals now have copper surfaces.

Although copper has its own unique color, it's very rarely seen like this in the ground. When it's exposed to air, it tarnishes to a vivid green. So what you see in the ground is the green "copper bloom."

It was a huge breakthrough when ancient people found that mixing copper with a little tin makes tough bronze. In a period called the Bronze Age, people made everything from knives and shields to cooking pots and statues from bronze.

COPPER: Group: Native element • Makeup: Copper • Color: Copper • Streak: Copper • Luster: Metallic • Hardness: 2.5 • SG: 8.9

DIAMOND AND SULFUR

HARDEST AND OLDEST

YELLOWEST AND SMELLIEST

DIAMOND AND SULFUR ARE EXTREMELY DIFFERENT BUT HAVE ONE BIG THING IN COMMON. Diamond is a hard, sparkly crystal and very, very rare. Although sulfur does occassionally form crystals, it's usually a dull yellow mass, made in lots of places. But diamond and sulfur are the only two nonmetal elements ever found naturally by themselves.

DIAMOND

Diamonds are the world's oldest gemstone, forged deep inside Earth, at least a billion years ago. They were carried to the surface in volcanic eruptions. Diamonds are the hardest natural substance on Earth. That's why diamonds are used in drill bits to bore through rock and to cut glass.

A diamond is usually clear, but has its own, unique internal sparkle or "fire" that makes it glitter. Probably the most famous diamond ever is the Hope Diamond. Owned by kings in France and England, it weighs 45.52 carats (.32 ounces or 9.1 g).

In 2004, scientists discovered a distant planet that might be mostly carbon and one-third pure diamond. They've also found a star that's essentially a diamond of ten billion trillion trillion carats!

DIAMOND: Group: Native element • Makeup: Carbon • Color: Colorless • Streak: White • Luster: Adamantine • Hardness: 10 • SG: 3.5

SULFUR

There's a lot of sulfur in oil and especially coal. Burning coal releases sulfur dioxide gas into the air. The gas joins with water to form sulfuric acid that then falls as acid rain. Acid rain can do terrible damage to forests and the soil.

Sulfur is bright yellow and burns. But don't ever try to light it—it gives off a poisonous gas and smells horrible! Sulfur was once called brimstone, or burning stone, and it was linked to the fires of hell! That may be partly because you can sometimes see crusts of it around hot volcanic springs and volcanic chimneys called fumaroles—as though smoke is coming from the underworld!

Mining sulfur is a stinky process, but sulfur is really useful for farm fertilizers, batteries, detergents, matches, and fireworks.

SULFUR: Group: Native element • Makeup: Sulfur • Color: Yellow • Streak: White or yellow • Luster: Glassy or earthy • Hardness: 2 • SG: 2

AMBER

SOME GEMS LOOK LIKE MINERALS, BUT AREN'T AT ALL. Amber, jet, and pearl are all made by living things. They're called mineraloids. Amber is the fossilized resin of ancient trees. The resin oozed down the bark to protect a tree from disease. This resin could trap small creatures and preserve them, perfectly, forever. Amber forms into smooth, irregular drops and nodules that can be polished to make soft stones. Meat-eating plants, ninety-nine-million-year-old ants, even dinosaur feathers have been found in amber!

AMBER: Group: Mineraloid • Makeup: Succinic acid • Color: Amber • Streak: White • Luster: Resinous • Hardness: 2 SG: • 1.1

JET

Jet started to form 180 million years ago, when flash floods swept Araucaria trees (like monkey puzzle trees) far out to sea. The logs became waterlogged and sank, and over millions of years turned deep black and hard. Jet looks a little like coal, but it formed in saltwater and doesn't contain so much carbon. That's why its streak is brown, not black. Jet's often been popular: The ancient Romans thought it was magical and could protect the wearer, including from snakes.

JET: Group: Mineraloid • Makeup: Carbon • Color: Black • Streak: Brown • Luster: Glassy • Hardness: 2 –2.5• SG: 1.1

PEARL

Pearls are actually mollusk mucus! They're made from the material nacre, or mother-of-pearl, which oysters and other shellfish use to line their shells. Pearls build up around a grain trapped in a shell. The mineral aragonite, drawn from the water around, and a protein called conchiolin glue grains together into a silver-white ball. Pearls grow in the wild and on pearl farms, where the farmer deliberately inserts an irritant into an oyster to make a pearl grow.

PEARL: Group: Mineraloid • Makeup: Pearl • Color: Pearly-white • Streak: White • Luster: Pearly • Hardness: 3.5–4 • SG: 2.9–3

SPACE ROCKS

ALL OUR NEAREST, NEIGHBORING PLANETS—Mercury, Mars, and Venus—are made from rock! So is our moon and the moons of other planets. Rocky asteroids orbit our sun in the asteroid belt, between Mars and Jupiter. Sometimes, they may fall through space and hit Earth as meteorites. Big meteorites hit the ground so hard that they instantly vaporize. But it's possible to find really tiny ones!

MERCURY HAS A HUGE IRON CORE.

THE CRUST OF VENUS SEEMS TO BE MOSTLY MADE OF BASALT.

Meteorites: Most meteorites are stony, made from silicates and other minerals, just like Earth is. Chondrite meteorites have little globules of olivine and pyroxene; achondrites look more like Earth's rocks. Iron meteorites are knobby lumps made mostly of iron and nickel. It's thought they come from the core of asteroids. One meteorite six miles (9.6 km) wide blasted down near the Yucatán Peninsula 66 million years ago. The debris cloaked the world in clouds and the dinosaurs died out pretty fast.

THERE'S A COVERING OF DUST THROWN UP BY METEORITE IMPACTS.

Earth's moon: Most of of our moon's rocks are igneous, formed by volcanic processes long ago. There are no sedimentary rocks, because there's no weather on the moon, and hardly any metamorphic rock, because everything stopped moving long ago.

THE MOON'S DARK, FLAT AREAS ARE BASALT.

MARS IS MOSTLY BASALT, COVERED WITH RED IRON DUST.

Mars: Recently, NASA has sent rovers to Mars, and they've photographed mudstones, sandstone, shale, and conglomerate. Sedimentary rocks on Mars means that, at some time, there must have been flowing water! Mars has the biggest volcano in the solar system: Olympus Mons.

GLOSSARY

ADAMANTINE Luster that is very brilliant and shiny.

BATHOLITH A large igneous intrusion shaped roughly like a dome, usually made of granite.

BED A layer of sedimentary rock.

BEDDING PLANE The boundary between one layer of sedimentary rock and another.

BIOCHEMICAL ROCKS Rocks such as limestone that are made from the remains of living things.

CARBONIFEROUS A period of geological history about 360–300 million years ago, when a lot of coal beds started to form.

CEMENTATION The gluing together of rock fragments by a paste of minerals, turning them into hard layers of sedimentary rock.

CHEMICAL ROCK A rock such as travertine that is made from fine powder left by dissolved minerals.

CLASTIC ROCK Rock such as sandstone and clay made from fragments or "clasts" of rock, such as sand, weathered out of other rocks.

COMPACTION When sediments are squeezed together, driving out water and air as more fragments drop on top.

COMPOUND A chemical combination of two or more elements.

CRETACEOUS A period of geological history 145–66 million years ago, when much chalk was formed.

CRYSTAL A solid that forms in a regular geometric shape.

DEPOSITION When fragments are dropped onto sea or lake beds, or are piled up on land by ice or wind.

DIKE A thin sheet of igneous intrusion, usually slanting or vertical.

ELEMENT A substance made from one of 120 or so basic kinds of atoms.

ERA A very long period of geological history, lasting hundreds of millions of years.

EROSION The wearing away of rocks by natural forces such as the weather, rivers, and waves.

EXTRUSIVE Describes igneous rock made from magma that erupts above the ground before going solid.

FELSIC Describes igneous rock that is light in color and forms from magma rich in silicates.

HABIT The shape in which a crystal tends to grow.

IGNEOUS Describes rock made from cooled magma.

INTRUSIVE Describes igneous rock made from magma that goes solid underground.

LAVA Hot, molten magma that gushes out of volcanoes.

LOPOLITH Igneous intrusion that runs in between rock layers to form a shallow bowl shape.

LUSTER The appearance of the surface of a mineral, depending on how it reflects light.

MAFIC Describes igneous rock that is dark in color and forms from magma low in silicates.

MAGMA Hot, molten rock that wells up from Earth's interior.

METAMORPHIC Describes rock made when other rocks are altered by extreme heat and pressure.

MINERAL One of thousands of different natural, solid crystal substances that make rocks.

NATIVE ELEMENT An element that occurs naturally in pure form in rocks, such as gold.

ORGANIC ROCK Rock made from the remains of living creatures.

PEGMATITE Very coarse-grained rocks formed from the last bits of magma to solidify.

PLUTON An igneous intrusion deep down.

PLUTONIC Describes rocks such as granite that solidified from magma deep underground.

PORPHYRIES Igneous rocks that have large grains embedded in much smaller grains.

SEDIMENT Material that settles on the bottom of the sea, a river, or a lake.

SEDIMENTARY ROCK Rock made in layers from sediment deposited on the seafloor and in other places.

SILICATE Mineral made from silicon and oxygen, usually with various other substances.

SILL A thin, usually horizontal sheet of igneous intrusion.

SPECIFIC GRAVITY A measurement comparing the density of a mineral to the density of water.

STRATA Layers of sedimentary rock.

STREAK The color of the mark a mineral leaves on a white tile.

VITREOUS Luster that looks like the luster of glass. Vitreous luster is also called glassy.

VOLCANIC ROCK Rock that forms from material erupted from volcanoes, including lava.

INDEX

Agate 97
Alabaster 61
Alloys 56
Almandine 111
Aluminum 51, 57, 80, 81, 90
Amber **118**
Amethyst 74, **95**
Andesite 19
Anthracite 37
Aragonite 32, 83, **84**, **85**, 119
Arsenic 52, 53
Azurite 51, 74, 87

Barite 51, **58**, **59**
Basalt 15, **20**, **21**
Batholiths 10, 11, 13
Bauxite 51, 57, **80**, **81**
Bedrock 21
Beryl **102**, **103**
Biotite 45, 115
Bronze 117

Calcite 32, 33, 43, **82**, **83**, 84
Calcium 51, 83
Carbon 37, 51
Carbon dioxide 83
Carbonates 51
Carnelian 97
Celestine 59
Cerium 63
Chalcedony **96**, **97**
Chalk **30**
Chert **31**
Chromium 62, 73, 103, 104
Cinnabar **55**, 74
Citrine 74, **94**
Clay 27
Clay ironstone 79
Coal **36**, **37**, 121

Conchiolin 85, 121
Contact metamorphism 39, 41
Copper 50, 57, 87, **116**, **117**
Core 11
Corundums 73
Crocoite **62**, 74
Crust 11
Crystals 48, 49, 101
Cuprite 57

Dacite 19
Diamonds 14, 15, 50, 104,
    120, **121**
Diatoms 89
Dikes 10, 11, 17
Dolerite **16**, **17**
Dolostone 33
Dripstone 35
Dunite 15
Dysprosium 56

Elbaite 101
Emerald 74, **104**, **105**

Feldspars 12, 45, 51, **90**, **91**
Firestone 109
Flint **31**
Fluorite **68**, **69**
Fossil fuels 37
Fossils 27, 30, 32, 33, 79

Gabbro 15
Gadolinium 57
Gahnite 71
Galena **54**, 56
Garnets 44, **110**, **111**
Gneiss **44**, **45**
Gold 9, 50, 74, **116**, **117**
Grains 7, 18, 21, 27, 94
Granite **12**, **13**

Gypsum 51, **60**, **61**

Habit 75
Halides 51
Halite 57, **66**, **67**
Hardness 77
Hematite 56, 57, **78**
Hornblende 45
Hornfels **40**, **41**

Iceland spar 83
Igneous rocks **6**, 10, 11, 19, 25,
    50, 88, 89
Iron 55, 56, 57, 78, 79

Jade **98**, **99**
Jadeite 99
Jasper 74
Jet **119**

Karst 33
Kimberlites 15
Kunzite 113

Labradorite 91
Lamprophyre 17
Lanthanum 63
Lapis lazuli 92, 93
Lava 10, 11, 18, 19, 20, 21,
    22, 23
Lazurite 74, **92**
Lead 50, 54, 56
Limestone 25, 31, **32**, **33**, 43,
    61, 82, 83
Limonite **79**
Lithium 17, 113
Lodestones 78
Lopoliths 10, 11, 15
Luster 75

Magma 10, 11, 13, 16, 17, 39, 41
Magnetite 56, 57, **78**
Malachite 51, 57, 74, **86, 87**
Mantle 11
Marble **42, 43**, 71
Mercury 55
Metamorphic rocks **7**, 38, 39, 50, 89
Meteorites **122, 123**
Mica 12, **114, 115**
Mineral chemistry 50, 51
Mineraloids 118
Minerals 5, 7, **8, 9**, 48, 49, 74, 75, 76, 77, 88
Mohs scale 77
Monazite 56, **63**
Moonstone 91
Morganite 103
Mudrocks 27
Mudstone 33, 45, 46
Muscovite 115

Nacre 85, 119
Native elements 50
Neodymium 56
Nephrite 99

Obsidian **22**, 31
Olivine 15, 74
Opal **106, 107**
Ores **56, 57**, 79
Orpiment **52, 53**, 74
Oxygen 51, 89

Paraiba 101
Pearls 85, **119**
Pegmatites 17, 49, 102, 112
Peridotite **14, 15**

Perlite 23
Petuntse 91
Phosphates 51
Phosphorus 51, 63
Plutons 11, 13
Porphyries 19
Primary oxides 51
Pumice 23
Pyrite **55**, 56, 74
Pyrope 111

Quartz 9, 12, 31, 45, 51, **94**, 95, 96

Rare earth elements 56, 63, 68, 113
Realgar **52, 53**
Regional metamorphism 39
Rhyolite **18, 19**, 23
Ring-stone 40
Rocks 4, 5, 6, 7
Ruby **72, 73**, 74, 104

Salt 66, 67
Salt lakes 49
Sand 88, 89, 94
Sandstone **28, 29**, 33, 61, 94
Sapphire 74, 104
Satin spar 61
Schist 45
Sediment pockets 49
Sedimentary rocks **6**, 25, 89
Selenite 61
Shale **26, 27**, 45
Siderite 57, **79**
Silica 19, 31, 89
Silicates 9, 51, **88, 89**, 90
Sills 10, 11, 16
Silt 27

Silver 50, 56
Slate **46, 47**
Smelting 56
Soapstone 115
Sodium 51, 57
Specific gravity (SG) 77
Spessartine 111
Sphene **112, 113**
Spinel **70, 71**
Spodumene **112, 113**
Stalactites 35, 83
Stalagmites 35, 83
Strata 25
Strontium 59
Sulfates 51
Sulfides 50
Sulfur 50, 53, 55, 74, 92, **120, 121**
Syenite 15

Talc 15, **114, 115**
Tectonic plates 45
Tiger's Eye **95**
Titanite 113
Topaz **108, 109**
Tourmaline **100, 101**
Travertine 25, **34**
Tufa 35
Tuff 23
Turquoise 51, **64, 65**

Vanadium 103, 104
Veins 49, 107, 109, 117
Vents 10
Volcanoes 10, 11, 14, 18, 19, 23
Vugs 102, 103

Wulfenite **63**

*Shiho Pate would like to thank
Zoe and Miriam for being the true gems in making this book!
To my strata-obsessed mom. And to my sister Momoko for
always being interested in my rock collection.*

Illustrated by Shiho Pate
Written by John Farndon
Consultant editor: Shannon Mahan, Research Geologist,
United States Geological Survey (USGS)
Art Director: Zoe Tucker

Library of Congress Cataloging-in-Publication Data Available

978-1-338-75367-7

10 9 8 7 6 5 4 3 2 1   22 23 24 25 26

Printed in China   38
First edition, October 2022